Inhibitory Synaptic Plasticity

Melanie A. Woodin · Arianna Maffei
Editors

Inhibitory Synaptic Plasticity

Springer

Editors
Melanie A. Woodin
Department of Cell and Systems Biology
Ramsay Wright Zoological Laboratory
University of Toronto
Toronto, Ontario
Canada
m.woodin@utoronto.ca

Arianna Maffei
Department of Neurobiology and Behavior
State University of New York Stony Brook
Stony Brook, New York
USA
Arianna.maffei@stonybrook.edu

ISBN 978-1-4419-6977-4 e-ISBN 978-1-4419-6978-1
DOI 10.1007/978-1-4419-6978-1
Springer New York Dordrecht Heidelberg London

Printed on acid-free paper

Springer is part of Springer Science+Business Media (www.springer.com)

Preface

Since its discovery and first publication in 1973, synaptic plasticity in its most popular form – the plasticity of excitatory glutamatergic synapses – has been widely investigated (Bliss and Lomo 1973; Malenka and Bear 2004). Every year hundreds of articles are published regarding the cellular, molecular and functional aspects of this fundamental form of plasticity. While it is widely accepted that excitatory plasticity is crucial for circuit refinement, learning, memory formation and storage, researchers now realize that this is not the only form of synaptic plasticity that these processes rely upon. A new idea that is slowly emerging is that multiple forms of synaptic and non-synaptic plasticity are involved in the formation, maintenance and modification of neural circuits, with all of the components acting in harmony.

A surprising discovery that has yet to be explored to its full potential is an unexpected, but very prominent form of synaptic plasticity: the plasticity of inhibitory GABAergic synapses. New findings are now accumulating that show inhibitory synaptic plasticity is widespread in neural circuits throughout the brain, and the functional roles of this form of plasticity are now being described.

We thought it timely to assemble a book that brings together these many new findings on inhibitory plasticity. Our goal was to effectively summarize and integrate studies that have examined the induction, expression and functional role of inhibitory synaptic plasticity at different organizational levels (from the neuron to the circuit). We soon realized that there are important common features, but also fundamentally different aspects of inhibitory GABAergic plasticity. Different areas of the brain require different induction paradigms, and different experimental approaches unveil a vast complexity of functional implications. In our minds the most effective way to convey all of this complexity was to divide the book into themes that highlight the major functions of inhibitory synaptic plasticity: Refinement, Addition, Pain and Injury, and Learning and Memory. This allowed us to explore in depth the cellular and molecular mechanisms underlying these various forms of inhibitory plasticity that appear to have only one thing in common: they occur at GABAergic synapses.

We brought together the new findings about the role of inhibitory synapse plasticity during postnatal circuit development and in the refinement of the circuit. In *Part 1: Refinement* Wang and Maffei discuss the role of inhibitory plasticity in the experience-dependent modulation of visual circuit excitability with a look on inhibitory circuits mediated by different subtypes of inhibitory neurons. Komatsu and Yoshimura bring together the known cellular mechanisms for inhibitory plasticity in neocortical inhibitory synapses and propose a role for this form of plasticity in visual cortical circuit development. Liu presents recent findings about the role of inhibitory plasticity in the processing of auditory stimuli, another sensory system that is strongly modulated by this form of plasticity. In addition he discusses his exciting new work about the role of inhibitory plasticity in adult rat mothers and their sensitivity to the calls of their pups. Kawaguchi and Hirano discuss a form of inhibitory GABAergic plasticity that is specific to cerebellar Purkinje neurons highlighting mechanisms of induction and expression with important similarities and relevant differences with those observed in other systems.

In *Part 2: Addiction* we brought together chapters discussing the role of inhibitory plasticity in two widespread forms of addiction: drugs and alcohol. Nugent brings to bear her knowledge of the mechanisms of inhibitory plasticity in the ventral tegmental area (VTA), an area of the brain that is involved in the general features of drug addiction. While the induction paradigm of this form of plasticity is identical to that used by Komatsu and Yoshimura in neocortex, the mechanisms of expression are profoundly different, highlighting the importance of specificity in the inhibitory circuit. Theile, Gonzalez and Morrisett present current thinking about the mechanisms of alcohol addiction and propose a new form of plasticity, an ethanol induced potentiation of GABA release in the VTA, as a possible mechanism to explain their new working hypothesis about the development of addiction.

Part 3: Pain and Injury reveals how serious insults to neurons and the nervous system results in drastic plasticity of inhibitory synapses, often reverting these synapses to a more immature developmental state. De Konnick and colleagues describe how neuropathic pain results from a loss of the chloride gradient that maintains inhibitory transmission. Timofeev et al. explain how inhibitory plasticity can results in deleterious consequences including seizure activity. And finally Moorhouse and Nabekura reveal the process whereby neuronal injury alters neuronal chloride transport, which in turn results in a loss of synaptic inhibition.

The plasticity induced during pain and injury results from pathological levels of activity. However inhibitory synaptic plasticity can also be induced by physiological levels of activity that occur during learning and memory. *Part 4: Learning and Memory* highlights how both activity itself, and neuromodulators activated by activity can modify the strength of inhibitory synapses. Balena, Acton and Woodin review literature which reveals how physiological levels of synaptic activity produce changes in chloride transport that alter $GABA_A$-mediated synaptic transmission. Chiu and Castillo tell of an elegant story of cross-talk between excitatory and inhibitory synapses. In this story glutamatergic activity facilitates endocannabinoid mobilization that causes a long-lasting reduction of GABA release, resulting in a heterosynaptic inhibitory long-term depression. But neuromodulation of inhibitory synapses isn't restricted to endocannabinoids. Gaiarsa and colleagues explain how neurotrophins have extended beyond their well known roles in mediating the survival, growth and differentiation of neurons to include short- and long-term modulations in the strength and development of GABAergic synapses.

As the reader moves through the themes and chapters of this book a complex landscape of mechanisms and functions emerges. Such richness and diversity clearly show how important the plasticity of inhibitory synapses is for circuit formation, refinement and ultimately for higher functions such as learning and memory. Inhibitory circuits do act as brakes that can stop or let go of specific kinds of information, sensory or cognitive, but they do so in a complex and refined manner that is specific to the many different types of inhibitory neurons that populate the brain (Klausberger and Somogyi 2008). To fully understand how brain functions develop and are maintained there is a need for studying the interaction between the different forms of plasticity. Bringing together a large portion of the current thinking about the plasticity of inhibitory synapses will hopefully foster new ideas and hypothesis for the study of brain circuits and their function.

Inhibitory synaptic plasticity is a relatively new field that has demonstrated exponential growth. Entering the term 'inhibitory synaptic plasticity' into a PubMed search yields 1,376 publications; with only 3% of those before the 90s, and over 70% in the last decade. With such fundamental advances being made at such a rapid rate it is exciting to imagine what the next decade of inhibitory synaptic plasticity research will reveal.

Toronto, Ontario, Canada Melanie A. Woodin
Stony Brook, NY, USA Arianna Maffei

References

Bliss TV, Lomo T (1973) Long-lasting potentiation of synaptic transmission in the dentate area of the anaesthetized rabbit following stimulation of the perforant path. J Physiol 232:331–356.
Klausberger T, Somogyi P (2008) Neuronal diversity and temporal dynamics: the unity of hippocampal circuit operations. Science 321:53–57.
Malenka RC, Bear MF (2004) LTP and LTD: an embarrassment of riches. Neuron 44:5–21.

Contents

Part I
Refinement

Chapter 1
The Many Faces of Inhibitory Plasticity: Adding Flexibility to Cortical Circuits Throughout Development

Lang Wang and Arianna Maffei

Abstract Neocortical circuits are highly interconnected networks of excitatory and inhibitory neurons. During postnatal development the connectivity and strength of excitatory and inhibitory synapses and the intrinsic properties of each neuron type sculpt the overall level of excitability of the circuit and support network function. Healthy neural circuits are characterized by a high sensitivity to changes in environmental stimuli and a finely tuned dynamic range. A balanced combination of excitatory and inhibitory inputs endows cortical circuits with the ability to maintain a dynamically stable level of excitability despite changes in sensory inputs. How this dynamically stable state is achieved during development is still matter of debate. In the past few decades, our knowledge of cortical neurogenesis, layer differentiation and circuit refinement has expanded dramatically. While most of the research has focused on the regulation of excitatory neocortical neurons, it is now accepted that inhibitory circuits contribute substantially to the achievement and maintenance of cortical circuit stability and function. Here we will focus on recent advancements in our understanding of the postnatal development of local inhibitory circuits and their role in the maintenance of cortical circuit excitability and stability.

Inhibitory Circuits

We will begin with a brief introduction about the many subtypes of inhibitory neurons in cortical circuits. Cortical inhibitory interneurons represent a morphologically, physiologically and chemically heterogeneous population of cells that perform different cortical functions (Burkhalter 2008). They have been classified in

A. Maffei (✉)
Department of Neurobiology and Behavior, State University of New York Stony Brook, Stony Brook, New York USA
e-mail: Arianna.maffei@stonybrook.edu

M.A. Woodin and A. Maffei (eds.), *Inhibitory Synaptic Plasticity*,
DOI 10.1007/978-1-4419-6978-1_1, © Springer Science+Business Media, LLC 2011

multiple ways. In rat cerebral cortex, three distinct subtypes of GABAergic interneurons have been identified based on the expression of calcium binding proteins: parvalbumin (PV), calretinin (CR) and somatostatin (SOM) (Gao et al. 2000; Gonchar et al. 2007). Inhibitory neurons have also been classified according to their distinct firing properties (Kawaguchi and Kubota 1997; Beierlein et al. 2003; Contreras 2004). Fast-spiking (FS) are identified by their high-frequency firing and the lack of frequency adaptation during depolarization. They include a morphologically diverse population composed of basket cells, chandelier cells and neurogliariform cells, which are also known as late-spiking (LS) cells (Gupta et al. 2000). Another prominent population of interneurons is composed of regular-spiking non pyramidal (RSNP) cells, which are subdivided into regularly adapting neurons and burst-spiking (BS) cells. BS cells are regular spiking inhibitory neurons that generate rebound burst spikes following hyperpolarization and are also known as low threshold spiking neurons (LTS; Beierlein et al. 2003). Interestingly, there are strong correlations between these classifications. For example, PV-expressing interneurons display FS properties (Xu et al. 2004; Butt et al. 2005), CR-expressing interneurons exhibit RS properties (Butt et al. 2005; Miyoshi et al. 2007), and SOM-expressing interneurons show BS properties (Butt et al. 2005; Miyoshi et al. 2007). Recent experiments suggest that the physiological subtype of cortical interneurons can be predicted by their temporal and spatial origins during neurogenesis (Butt et al. 2005).

Interneuron-Specific Synaptic Connections

Diverse subtypes of cortical interneurons form distinct synaptic contacts on pyramidal neurons and non-pyramidal neurons (Thomson and Lamy 2007). PV-expressing basket cells preferentially innervate somata and proximal dendrites of pyramidal neurons, while PV-expressing chandelier cells are thought to form connections with axon initial segments of pyramidal neurons (Kawaguchi and Kubota 1997; Gonchar and Burkhalter 1999b; Woodruff and Yuste 2008). Since the axon initial segment is the site where action potentials are generated, the synaptic contacts in this region are in a privileged position to control the output of pyramidal neurons. Chandelier cells were initially thought to be the only type of non-pyramidal cell to perform this fundamental function, but according to recent results the axon terminals of SOM-expressing cells also form symmetric synapses with the axon initial segments of pyramidal neurons in visual cortical supragranular layers (Gonchar et al. 2002). Unlike chandelier cells that innervate exclusively the axon initial segments of pyramidal neurons, SOM neurons also make synaptic contacts onto somata and dendrites of pyramidal neurons (Kawaguchi and Kubota 1997). This anatomical organization suggests that SOM neurons may modulate both the integration of inputs and the output of pyramidal neurons. All of these results imply a more complex regulation of pyramidal neuron output arising from the integration of different sources of inhibitory inputs.

Two distinct subtypes of interneurons, PV positive interneurons and those containing the neuropeptide cholecystokinin (CCK; Freund 2003) often form perisomatic nets consisting of multiple terminals with large boutons clustered around pyramidal soma and proximal dendrites (Wang et al. 2002; Chattopadhyaya et al. 2004). These large webs of perisomatic inhibitory synapses may be exceptionally suited for controlling the output of large groups of pyramidal neurons via synchronization of action potential firing. The formation and maintenance of terminal branches appears to be independent of neuronal activity and possibly relies on intrinsic developmental cues. Activity is necessary for the proliferation and extension of boutons on the perisomatic contacts (Chattopadhyaya et al. 2004). Thus the formation of perisomatic innervations likely depends on intrinsic developmental cues, whereas its maturation is modulated by driving inputs. CR-expressing neurons predominantly innervate other non-pyramidal interneurons (Gonchar and Burkhalter 1999a).

Maturation of Inhibitory Circuits

Neurotrophic factors including nerve growth factor (NGF), brain-derived neurotrophic factor (BDNF), and neurotrophin-3/4 (NT-3/4) are implicated in the regulation of neuronal differentiation (Ghosh and Greenberg 1995), axonal and dendritic growth (Horch and Katz 2002; Spatkowski and Schilling 2003), and synapse formation (Genoud et al. 2004; Gomes et al. 2006). In 1997, Marty and collaborators proposed that neurotrophins may play a role in cortical development and plasticity by regulating GABAergic inhibitory interneurons (Marty et al. 1997). Subsequent experiments confirmed this hypothesis. In cortical cell culture, BDNF promotes interneuron axon growth, stimulates the expression of GABA and of several calcium-binding proteins, and regulates the strength of synaptic inhibition (Cellerino 1996; Rutherford 1998). In transgenic mice in which the postnatal increase in BDNF in the forebrain was accelerated, the maturation of the GABAergic inhibitory circuit was completed earlier than in wild types, suggesting a causal relationship between the levels of BDNF and the maturation of inhibition (Huang et al. 1999). Another factor crucial to the maturation of cortical inhibition is polysialic acid (PSA), a long, linear homopolymer of a-2,8-linked sialic acid attached almost exclusively to the neural cell adhesion molecule (NCAM) in vertebrates (Di Cristo et al. 2007). During postnatal development, premature decline of PSA levels in visual cortex results in enhanced inhibitory synaptic transmission and accelerated maturation of perisomatic innervations by basket interneurons (Di Cristo et al. 2007). Recently a new factor was identified as fundamental for the postnatal maturation of inhibitory synaptic transmission: the homeodomain protein Otx2. Otx2 is normally required for embryonic head formation (Acampora et al. 2001), and is also restricted to relay centers along the primary visual pathway at birth, including retina, lateral geniculate nucleus (LGN) and visual cortex (Nothias et al. 1998). The expression of Otx2 in regions of the dorsal forebrain in rats was expected to be lost by postnatal days 13 (P13) until very recently, when Hensch and collaborators found non-cell-autonomous Otx2 coordinates the maturation of PV-expressing interneurons even after the third postnatal week and that visual experience promotes the accumulation of Otx2 in PV-cells (Sugiyama et al. 2008). The implications of these results are quite important: the experience–dependent transfer of a homeoprotein may regulate the maturation of inhibitory circuits during the first few weeks of postnatal life.

Cortical Inputs to Inhibitory Interneuron

The function of an individual neuron is largely determined by the synaptic input onto that neuron. Differently from inhibitory innervations in areas without "laminar patterns", cortical interneurons receive distinct laminar inputs that are subtype-specific (Thomson and Lamy 2007). In layer 2/3 of rat visual cortex, FS basket cells receive strong excitatory inputs from middle cortical layers (Dantzker and Callaway 2000). In contrast, adapting inhibitory interneurons receive their strongest excitatory inputs either vertically from deep layers or laterally from within layer 2/3 (Dantzker and Callaway 2000). In layer 5 of rat frontal cortex, it was found that the probability of reciprocal connectivity between pyramidal neurons and FS cells within the same layer was much higher than that between pyramidal neurons and non-FS cells, whereas non-FS cells in layer 5 received a more substantial set of inputs from pyramidal neurons in layer 2/3 (Cruikshank et al. 2007; Otsuka and Kawaguchi 2009). Such layer- and subtype-specific inputs contribute to the functional diversity of cortical inhibitory circuits and might depend on the fact that cortical circuit maturation is driven simultaneously by the cooperation of activity-independent cues within the local environment and by sensory inputs. Distinct subtypes of interneurons in layer 4 have specific laminar input (Hajós et al. 1997; Beierlein et al. 2003) but little is known about how their intracortical input integrates with that carried by thalamocortical afferents. Both FS and RSNP inhibitory neurons are thought to take part in a feedforward inhibitory circuit that receives thalamic input as well as in the widespread feedback recurrent inhibitory network (Beierlein et al. 2003; Cruikshank et al. 2007; Thomson and Lamy 2007; Tan et al. 2008; Hull et al. 2009).

Cortical Circuit Refinement

As early as in 1963, Sperry, Hubel and Wiesel had noted that the high degree of anatomical and functional precision in the visual pathway is present even before vision begins (Sperry 1963; Wiesel and Hubel 1963a). According to these findings, it was hypothesized that the wiring of cortical circuits relies on "innate cues" (Sperry 1963; Wiesel and Hubel 1963b). After eye opening, however, the postnatal development of cortical circuits occurs through a sequence of adjustments in connectivity and synaptic strength that promote the stabilization of active synaptic pathways and the elimination of synapses that are not stimulated by the appropriate input (Katz and Shatz 1996). This fine scale refinement occurs during periods of heightened sensitivity to changes in the driving input known as "critical" periods (Hensch 2004). About 50 years ago, Hubel and Wiesel discovered that the reduction or the elimination of visual input to one eye during the first few weeks of postnatal development induces a dramatic anatomic reorganization of the primary visual cortex (Wiesel and Hubel 1965, 1974). Similar reorganizations were subsequently observed in somatosensory (Inan 2007), auditory (Yu 2007), in motor cortex (Harms 2008), olfactory bulb (Tyler 2007; Marks 2006) and in the superior colliculus (Chandrasekaran 2007). Critical periods usually last only a few weeks; if sensory inputs are manipulated after the expected closing of these windows of opportunity, the rearrangements observed are not as dramatic or do not occur at all (Fagiolini et al. 1994; Hensch 2004). The opening and duration of the windows for plasticity vary depending on the area of cortex and in some brain areas more than one critical period has been identified (Feller and Scanziani 2005; Tagawa et al. 2005; Hensch 2004). The study of critical periods is especially important to understand basic mechanisms controlling neural circuit stability and flexibility in the face of the constant changes in environmental inputs. In the following paragraphs we will discuss the potential roles for inhibitory circuits and their plasticity in acquiring and maintaining circuit stability during critical periods for visual cortical development.

Plasticity of Inhibitory Synapses onto Cortical Pyramidal Neurons

Inhibitory circuits show complex, broadly defined plasticity: the relative proportion of specific populations of neurons can be regulated differentially in areas of the cortex processing different functions and/or in response to exposure to environmental factors (Harvey 2001; Fountain 2000); the axonal arborization and complexity can change both during development and in the adult in an activity-dependent manner (Dantzker and Callaway 1998; Chattopadhyaya et al. 2004); the subunit composition of $GABA_A$ receptors is developmentally regulated (Dunning et al. 1999; Heinen et al. 2004). Finally, the strength of their functional synapses and their intrinsic properties are modulated by activity (Maffei 2006; Sun 2009). The diversity of interneurons populations and the complexity of their plasticity allows for an extraordinary flexibility in the way inhibitory neurons regulate cortical circuit excitability. During development, a subset of inhibitory synapses onto pyramidal neurons mediated by specific interneuron types may change the subunit composition of $GABA_A$ receptors on the postsynaptic terminal (Dunning et al. 1999; Heinen et al. 2004). $GABA_A$ receptors from most basket neurons contain alpha 3 and alpha 5 early in development and switch to alpha 1 around the third postnatal week. This switch is considered a hallmark of the maturation of GABAergic synapses

(Dunning 1999; Heinen 2004; Bosman et al. 2002). Differently, inhibitory synapses postsynaptic to chandelier cells express alpha 2 containing $GABA_A$ receptors throughout development and those postsynaptic to bipolar neurons contain mostly alpha 5 subunits (Ali and Thomson 2008). The difference in subunit composition underlies specific pharmacological and functional properties. In visual cortex, lack of alpha 1 alters the opening and duration of the classical critical period for ocular dominance plasticity (Hensch et al. 1998). Differently, lack of alpha 2 subunits during development strongly affects pyramidal neuron input/output function, but does not change neuronal responsiveness to visual stimulation (Fagiolini and Hensch 2000). Alpha 5 containing synapses mediated by bipolar neurons are thought to control the dendritic integration of inputs onto excitatory neurons (Ali and Thomson 2008). Morphologically, the structures of the axonal arbors of inhibitory neurons are regulated in an activity dependent manner (Chattopadhyaya et al. 2004). The reorganization of inhibitory synaptic connectivity depends on the maturation and plasticity of functional synapses. An interesting finding is that somatic inhibitory synapses have the ability to respond to changes in driving input with opposite direction plasticity during two stages of development: the pre-critical period (from eye opening to the end of the third postnatal week) and the critical period for visual cortical plasticity (Maffei et al. 2004, 2006; Maffei and Turrigiano 2008b). While the specific mechanisms of these forms of plasticity are yet to be elucidated, the timing of the change in the direction of inhibitory plasticity raises the possibility that cellular mechanisms underlying the maturation of inhibitory synapses might also promote the switch in direction of their plasticity. A variety of patterns of activity producing depression and potentiation have been reported for fast inhibitory synapses (Komatsu and Iwakiri 1993; Holmgren and Zilberter 2001; Nugent et al. 2007). Potentiation of inhibitory inputs onto cortical pyramidal neurons transmission was induced with extracellular high frequency stimulation (Komatsu and Iwakiri 1993). This groundbreaking work demonstrated for the first time that inhibitory synapses are plastic and suggested this plasticity as a potential new mechanism for the formation of memory. High frequency potentiation of IPSPs depends on the heterosynaptic activation of NMDA receptors (Komatsu and Iwakiri 1993; Kurotani et al. 2008). In layer 5 of rodent visual cortex high frequency stimulation also induced a form $GABA_B$ receptor dependent inhibitory LTP (Komatsu 1996). Both $GABA_B$ and NMDA dependent forms of inhibitory plasticity relied on postsynaptic dendritic exocytosis, possibly promoting the insertion of new $GABA_A$ receptors in the postsynaptic membrane (Komatsu and Iwakiri 1993; Komatsu 1996). Extracellular stimulation does not allow the subtype of inhibitory interneuron mediating plasticity to be identified, and simultaneously activates quite strongly all of the axons (excitatory and inhibitory) travelling in the region of stimulation, suggesting that the NMDA dependence of inhibitory synaptic plasticity might be specific to this particular mode of induction. Inhibitory circuits mediated by different interneuron subtypes are not all affected by experience in the same way (Maffei et al. 2004; Bartley et al. 2008). The diversity in experience dependent plasticity is also supported by growing experimental evidence in favor of a rich set of plastic and experience-dependent changes in inhibitory circuits with potentially important implications for circuit stability and function (Komatsu 1996; Holmgren and Zilberter 2001; Fritschy and Brünig 2003; Haas et al. 2006; Bartley et al. 2008; Maffei and Fontanini 2009).

Specific Inhibitory Circuits and Their Long Term Plasticity

The use of paired recording electrophysiological techniques and of transgenic mouse lines in which the different interneuron subtypes can be identified offered new possibilities to investigate the specificity of inhibitory plasticity. The first fundamental finding of this novel approach was that plasticity is induced at specific connections of inhibitory onto excitatory neurons with different patterns of activity depending on the subtype of inhibitory neuron, the brain circuit under analysis, and development (Jiao et al. 2006; Maffei and Turrigiano 2008b). In layer 2/3 of neocortex, pyramidal neuron bursting followed by delayed FS neuron spiking potentiates or depresses inhibitory synapses depending on the duration of the delay (Holmgren and Zilberter 2001). Both potentiation and depression of FS to pyramidal neuron inhibitory plasticity depends on calcium influx into the postsynaptic terminal and is independent of $GABA_B$ receptor activation (Komatsu 1996; Holmgren and Zilberter 2001). The requirements for the induction of FS mediated plasticity in layer 4 are different than those observed in layer 2/3. Successful potentiation of these synapses depends on a timed activation of pre (FS) and postsynaptic (Pyramidal) neurons but FS interneuron bursting needs to be paired with pyramidal neuron subthreshold depolarization (Maffei et al. 2006). Despite these differences in induction, the LTP of inhibition in layer 2/3 and 4 present some interesting similarities: both appear to have a postsynaptic site of expression and no detectable changes at the presynaptic site (Holmgren and Zilberter 2001; Maffei et al. 2006). Several intracellular pathways are involved in regulating the number of $GABA_A$ receptors at inhibitory synapses (Heuschneider and Schwartz 1989; Brandon et al. 2000; Kumar et al. 2005; Bogdanov et al. 2006) but whether they are also involved in the fast transport required for plasticity is not known. Much less is known about the patterns of activity inducing plasticity at inhibitory synapses mediated by non-FS interneurons. Experiments measuring the experience-dependent changes at low threshold spiking interneurons (LTS) in the barrel cortex and at regular spiking non pyramidal neurons (RSNP) in visual cortex prove that these inhibitory connections onto pyramidal neurons are plastic and sensitive to changes in the environment (Maffei et al. 2004; Bartley et al. 2008). These connections present some interesting differences between cortical areas. In visual cortex, reduction of visual drive right at eye opening strengthens their synapses onto pyramidal neurons, decreases their connection probability, and leaves their short term dynamics unaffected (Maffei et al. 2004). In barrel cortex, LTS neurons – also classified as SOM neurons – change their short term dynamic in response to activity blockade (Bartley et al. 2008). In slice culture from the somatosensory cortex, 5 days incubation with tetrodotoxin, a treatment that completely silences neurons in the circuit, specifically affected the short term dynamics of synapses mediated by LTS neurons onto pyramidal neurons, leaving FS driven inhibition unaffected (Bartley et al. 2008). These changes may shift the relative influence of the two most prevalent inhibitory circuits, differently in different cortical areas, and may be specific to the function of RSNP and LTS neurons perform in their specific circuits. Combinations of different forms of inhibitory plasticity contribute to sculpting the level of excitability of the circuit and the integration of sensory inputs (Bartley et al. 2008; Maffei and Fontanini 2009).

Functional Implications

What is the role of inhibitory plasticity in cortical circuits? The ability to directly control pyramidal neuron input integration, excitability, and output suggest that inhibitory synaptic transmission is fundamental for cortical circuit function. Furthermore, the sensitivity of inhibitory synapses to changes in input and their ability to adjust their strength in response to specific patterns of activity allows fast regulation of cortical circuit activity and increases flexibility (Maffei and Fontanini 2009). During the first few weeks of postnatal development, the maturation of GABAergic transmission promotes the maturation of connectivity and excitability of local microcircuits (Komatsu 1994; Hensch and Fagiolini 2005; Kotak et al. 2008; Maffei and Turrigiano 2008a). Part of this process is activity dependent and driven by the incoming input (Jiao et al. 2006; Katagiri et al. 2007; Kotak et al. 2008; Maffei and Turrigiano 2008a). Inputs from the environment together with the state of excitability of the microcircuit, in turn, promote the refinement of cortical maps (Cang et al. 2005; Chandrasekaran et al. 2007; Cheetham et al. 2007). While the process of cortical map refinement has been ascribed for the most part to changes in synaptic weight and connectivity of glutamatergic synapses (Bender et al. 2006; Cheetham et al. 2008; Yoon et al. 2009), there is now growing evidence that inhibitory circuits and their plasticity are also prominently involved (Hensch et al. 1998; Maffei et al. 2004, 2006; Maffei and Turrigiano 2008a; Mainardi et al. 2009). A strong correlation between the refinement of cortical maps and inhibitory synaptic transmission and plasticity lies in the strong temporal correlation between maturation of inhibitory circuits and that of cortical connectivity (Fagiolini and Hensch 2000; Maffei et al. 2004, 2006; Katagiri et al. 2007). Specific inhibitory circuits may contribute differently to the refinement process, as suggested by the differential effects of sensory deprivation paradigms on the two major populations of inhibitory neurons (Maffei et al. 2004; Bartley et al. 2008; Sun 2009). In rodent neocortex, both the barrel field of somatosensory cortex and primary visual cortex show depression of FS synaptic inhibition onto pyramidal neurons in response to sensory deprivation during early postnatal development (Maffei et al. 2004; Jiao et al. 2006). The decrease in strength of somatic inhibition has been observed in layer 4, the main recipient of thalamocortical projections. In layer 4, a substantial remodeling of FS perisomatic axonal arbors around pyramidal neurons was also reported (Chattopadhyaya et al. 2004). In layer 2/3, dark-rearing from birth, a paradigm that delays the development of visual cortex, reduces the number of somatic puncta from GAD65-positive inhibitory terminals onto pyramidal neurons (Kreczko et al. 2009), but whether this is also accompanied by a decrease in strength of functional synapses is currently unknown. Together these results suggest a strong experience-dependence of FS inhibitory synaptic strength and connectivity. In visual cortex, one of the possible interpretations of the reduced FS inhibition is that it preserves the overall state of excitability of the circuit in the face of a reduced driving input, thus promoting a homeostatic regulation of cortical excitability (Maffei and Fontanini 2009). In favor of this hypothesis, monocular deprivation does not shift ocular dominance between eye opening and the beginning of the classical critical period for amblyopia (Fagiolini et al. 1994). The lack of ocular dominance plasticity might depend on the increase in cortical excitability (Maffei et al. 2004). In the barrel cortex, the decrease in somatic inhibition might play a similar homeostatic role (Jiao et al. 2006; Bartley et al. 2008). After the third postnatal week, visual deprivation was reported to potentiate FS mediated inhibition onto pyramidal neurons (Maffei et al. 2006). The switch in sign of FS inhibitory plasticity correlates with the expected time of initiation of the critical period for ocular dominance plasticity (Fagiolini et al. 1994). An intriguing possibility is that the potentiation of inhibition contributes to the silencing of neurons driven by the deprived eye, thus promoting the shift in ocular dominance. So far, a switch in sign of inhibitory plasticity was reported only for visual cortex, but might be a general mechanism for cortical map refinement. Beyond cortical maps, inhibitory synaptic transmission located in very critical compartments of pyramidal neurons might contribute to shaping neuronal receptive fields. There is in fact evidence for a developmentally regulated sharpening of cortical receptive fields that is temporally correlated with the maturation of inhibitory synapses, and for the role of lateral inhibition in shaping the size and morphology of receptive fields (de la Rocha et al. 2008) (Fig. 1.1).

Conclusion

A growing body of evidence suggests that inhibitory synaptic plasticity is involved in regulating neuronal function, cortical circuit connectivity, and sensory input integration. The richness in plasticity and the specificity of inhibitory circuits might provide fast and efficient regulation of local circuit excitability, which, in turn, will affect the integration of pyramidal neurons input/output function, circuit stability and network computation. Beyond sensory cortices, GABAergic plasticity has been observed in areas involved in several learning and emotional functions, suggesting that inhibitory transmission and plasticity play complex roles in cortical circuit function.

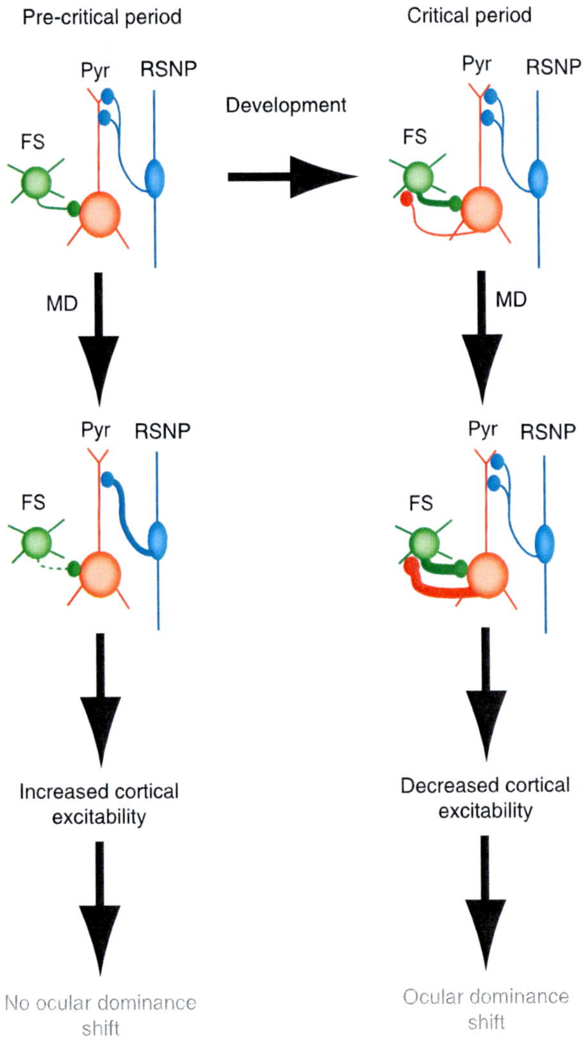

Fig. 1.1 Developmental regulation of inhibitory plasticity and it functional implications. The cartoon represents a summary of the data describing developmental changes in inhibitory synaptic strength, their experience dependent changes in response to alteration of sensory input, their effects on circuit excitability and their possible functional implications. Pre-critical period is defined in rodents as the postnatal week between eye opening (postnatal day 14, P14) and the expected beginning of the critical period for visual cortical plasticity (P21). The critical period is expected to begin at P21 and to end by P35. Pyr: pyramidal neurons; FS: fast spiking inhibitory interneurons; RSNP: regular spiking inhibitory interneurons. The thickness of the lines indicates the strength of the connections (thicker: stronger; dotted: weaker)

Acknowledgments This work was funded by the NIH/NEI R01 grant EY019885 (AM). We thank Martha Stone and Alfredo Fontanini for useful comments and discussions.

References

Acampora D, Gulisano M, Broccoli V, Simeone A (2001) Otx genes in brain morphogenesis. Prog Neurobiol 64:69–95.

Ali A, Thomson A (2008) Synaptic alpha 5 subunit-containing GABAA receptors mediate IPSPs elicited by dendrite-preferring cells in rat neocortex. Cerebral Cortex 18:1260–1271.

Bartley AF, Huang ZJ, Huber KM, Gibson JR (2008) Differential activity-dependent, homeostatic plasticity of two neocortical inhibitory circuits. J Neurophysiol 100:1983–1994.

Beierlein M, Gibson J, Connors B (2003) Two dynamically distinct inhibitory networks in layer 4 of the neocortex. J Neurophysiol 90:2987–3000.

Bender K, Allen C, Bender V, Feldman D (2006) Synaptic basis for whisker deprivation-induced synaptic depression in rat somatosensory cortex. J Neurosci 26:4155–4165.

Bogdanov Y, Michels G, Armstrong-Gold C, Haydon P, Lindstrom J, Pangalos M, Moss S (2006) Synaptic GABAA receptors are directly recruited from their extrasynaptic counterparts. EMBO J 25:4381–4389.

Bosman L, Rosahl T, Brussaard A (2002) Neonatal development of the rat visual cortex: synaptic function of GABAA receptor alpha subunits. J Physiol 545:169–181.

Brandon N, Delmas P, Kittler J, McDonald B, Sieghart W, Brown D, Smart T, Moss S (2000) GABAA receptor phosphorylation and functional modulation in cortical neurons by a protein kinase C-dependent pathway. J Biol Chem 275:38856–38862.

Burkhalter A (2008) Many specialists for suppressing cortical excitation. Front Neurosci 2:155–167.

Butt SJ, Fuccillo M, Nery S, Noctor S, Kriegstein A, Corbin JG, Fishell G (2005) The temporal and spatial origins of cortical interneurons predict their physiological subtype. Neuron 48:591–604.

Cang J, Rentería R, Kaneko M, Liu X, Copenhagen D, Stryker M (2005) Development of precise maps in visual cortex requires patterned spontaneous activity in the retina. Neuron 48:797–789.

Cellerino A, Maffei L (1996) The action of neurotrophins in the development and plasticity of the visual cortex. Prog Neurobiol 49:53–71.

Chandrasekaran A, Shah R, Crair M (2007) Developmental homeostasis of mouse retinocollicular synapses. J Neurosci 27:1746–1755.

Chattopadhyaya B, Di Cristo G, Higashiyama H, Knott GW, Kuhlman SJ, Welker E, Huang ZJ (2004) Experience and activity-dependent maturation of perisomatic GABAergic innervation in primary visual cortex during a postnatal critical period. J Neurosci 24:9598–9611.

Cheetham C, Hammond M, Edwards C, Finnerty G (2007) Sensory experience alters cortical connectivity and synaptic function site specifically. J Neurosci 27:3456–3465.

Cheetham C, Hammond M, McFarlane R, Finnerty G (2008) Altered sensory experience induces targeted rewiring of local excitatory connections in mature neocortex. J Neurosci 28:9249–9260.

Contreras D (2004) Electrophysiological classes of neocortical neurons. Neural Netw 17:633–646.

Cruikshank S, Lewis T, Connors B (2007) Synaptic basis for intense thalamocortical activation of feedforward inhibitory cells in neocortex. Nat Neurosci 10:462–468.

Dantzker JL, Callaway EM (1998) The development of local, layer-specific visual cortical axons in the absence of extrinsic influences and intrinsic activity. J Neurosci 18:4145–4154.

Dantzker JL, Callaway EM (2000) Laminar sources of synaptic input to cortical inhibitory interneurons and pyramidal neurons. Nat Neurosci 3:701–707.

Dunning D, Hoover C, Soltesz I, Smith M, O'Dowd D (1999) GABA(A) receptor-mediated miniature postsynaptic currents and alpha-subunit expression in developing cortical neurons. Jounal of Neurophysiology 82:3286–3297.

de la Rocha J, Marchetti C, Schiff M, Reyes AD (2008) Linking the response properties of cells in auditory cortex with network architecture: cotuning versus lateral inhibition. J Neurosci 28:9151–9163.

Di Cristo G, Chattopadhyaya B, Kuhlman SJ, Fu Y, Belanger MC, Wu CZ, Rutishauser U, Maffei L, Huang ZJ (2007) Activity-dependent PSA expression regulates inhibitory maturation and onset of critical period plasticity. Nat Neurosci 10:1569–1577.

Dunning D, Hoover C, Soltesz I, Smith M, O'Dowd D (1999) GABA(A) receptor-mediated miniature postsynaptic currents and alpha-subunit expression in developing cortical neurons. J Neurophysiol 82:3286–3297.

Fagiolini M, Hensch T (2000) Inhibitory threshold for critical-period activation in primary visual cortex. Nature 404:183–186.

Fagiolini M, Pizzorusso T, Berardi N, Domenici L, Maffei L (1994) Functional postnatal development of the rat primary visual cortex and the role of visual experience: dark rearing and monocular deprivation. Vision Res 34:709–720.

Feller M, Scanziani M (2005) A precritical period for plasticity in visual cortex. Curr Opin Neurobiol 15:94–100.

Fountain N (2000) Status epilepticus: risk factors and complications. Epilepsia 41:23–30.

Freund TF (2003) Interneuron Diversity series: Rhythm and mood in perisomatic inhibition. Trends Neurosci 26:489–495.

Fritschy J, Brünig I (2003) Formation and plasticity of GABAergic synapses: physiological mechanisms and pathophysiological implications. Pharmacol Ther 98:299–323.

Gao WJ, Wormington AB, Newman DE, Pallas SL (2000) Development of inhibitory circuitry in visual and auditory cortex of postnatal ferrets: immunocytochemical localization of calbindin- and parvalbumin-containing neurons. J Comp Neurol 422:140–157.

Genoud C, Knott GW, Sakata K, Lu B, Welker E (2004) Altered synapse formation in the adult somatosensory cortex of brain-derived neurotrophic factor heterozygote mice. J Neurosci 24:2394–2400.

Ghosh A, Greenberg ME (1995) Distinct roles for bFGF and NT-3 in the regulation of cortical neurogenesis. Neuron 15:89–103.

Gomes RA, Hampton C, El-Sabeawy F, Sabo SL, McAllister AK (2006) The dynamic distribution of TrkB receptors before, during, and after synapse formation between cortical neurons. J Neurosci 26:11487–11500.

Gonchar Y, Burkhalter A (1999a) Connectivity of GABAergic calretinin-immunoreactive neurons in rat primary visual cortex. Cereb Cortex 9:683–696.

Gonchar Y, Burkhalter A (1999b) Differential subcellular localization of forward and feedback interareal inputs to parvalbumin expressing GABAergic neurons in rat visual cortex. J Comp Neurol 406:346–360.

Gonchar Y, Wang Q, Burkhalter A (2007) Multiple distinct subtypes of GABAergic neurons in mouse visual cortex identified by triple immunostaining. Front Neuroanat 1:3.

Gonchar Y, Turney S, Price JL, Burkhalter A (2002) Axo-axonic synapses formed by somatostatin-expressing GABAergic neurons in rat and monkey visual cortex. J Comp Neurol 443:1–14.

Gupta A, Wang Y, Markram H (2000) Organizing principles for a diversity of GABAergic interneurons and synapses in the neocortex. Science 287:273–278.

Haas J, Nowotny T, Abarbanel H (2006) Spike-timing-dependent plasticity of inhibitory synapses in the entorhinal cortex. J Neurophysiol 96:3305–3313.

Hajós F, Staiger J, Halasy K, Freund T, K Z (1997) Geniculo-cortical afferents form synaptic contacts with vasoactive intestinal polypeptide (VIP) immunoreactive neurons of the rat visual cortex. Neuroscience Letters 228:179–182.

Harms K, Rioult-Pedotti M, Carter D, Dunaevsky A (2008) Transient spine expansion and learning-induced plasticity in layer 1 primary motor cortex. J Neurosci 28:5686–5690.

Harvey J, Romano A, Gabriel M, Simansky K, Du W, Aloyo V, Friedman E (2001) Effects of prenatal exposure to cocaine on the developing brain: anatomical, chemical, physiological and behavioral consequences. Neurotox Res 3:117–143.

Heinen K, Bosman L, Spijker S, van Pelt J, Smit A, Voorn P, Baker R, Brussaard A (2004) GABAA receptor maturation in relation to eye opening in the rat visual cortex. Neuroscience 124:161–171.

Hensch T (2004) Critical period regulation. Ann Rev Neurosci 27:549–579.

Hensch T, Fagiolini M (2005) Excitatory-inhibitory balance and critical period plasticity in developing visual cortex. Prog Brain Res 147:115–124.

Hensch T, Fagiolini M, Mataga N, Stryker M, Baekkeskov S, Kash S (1998) Local GABA circuit control of experience-dependent plasticity in developing visual cortex. Science 282:1504–1508.

Heuschneider G, Schwartz R (1989) cAMP and forskolin decrease gamma-aminobutyric acid-gated chloride flux in rat brain synaptoneurosomes. Proc Natl Acad Sci 86:2938–2942.

Holmgren CD, Zilberter Y (2001) Coincident spiking activity induces long-term changes in inhibition of neocortical pyramidal cells. J Neurosci 21:8270–8277.

Horch HW, Katz LC (2002) BDNF release from single cells elicits local dendritic growth in nearby neurons. Nat Neurosci 5:1177–1184.

Huang ZJ, Kirkwood A, Pizzorusso T, Porciatti V, Morales B, Bear MF, Maffei L, Tonegawa S (1999) BDNF regulates the maturation of inhibition and the critical period of plasticity in mouse visual cortex. Cell 98:739–755.

Hull C, Isaacson J, Scanziani M (2009) Postsynaptic mechanisms govern the differential excitation of cortical neurons by thalamic inputs. J Neurosci 29:9127–9136.

Inan M, Crair M (2007) Development of cortical maps: perspectives from the barrel cortex. Neurosci 13:49–61.

Jiao Y, Zhang C, Yanagawa Y, Sun Q (2006) Major Effects of Sensory Experiences on the Neocortical Inhibitory Circuits. J Neurosci 26:8691–8701.

Katagiri H, Fagiolini M, Hensch T (2007) Optimization of somatic inhibition at critical period onset in mouse visual cortex. Neuron 53:805–812.

Katz L, Shatz C (1996) Synaptic activity and the construction of cortical circuits. Science 274:1133–1138.

Kawaguchi Y, Kubota Y (1997) GABAergic cell subtypes and their synaptic connections in rat frontal cortex. Cereb Cortex 7:476–486.

Komatsu Y (1994) Age-dependent long-term potentiation of inhibitory synaptic transmission in rat visual cortex. J Neurosci 14:6488–6499.

Komatsu Y (1996) GABAB receptors, monoamine receptors, and postsynaptic inositol trisphosphate-induced Ca2+ release are involved in the induction of long-term potentiation at visual cortical inhibitory synapses. J Neurosci 16:6342–6352.

Komatsu Y, Iwakiri M (1993) Long-term modification of inhibitory synaptic transmission in developing visual cortex. Neuroreport 4:907–910.

Kotak V, Takesian A, Sanes D (2008) Hearing loss prevents the maturation of GABAergic transmission in the auditory cortex. Cerebral Cortex 18:2098–2108.

Kreczko A, Goel A, Song L, Lee HK (2009) Visual deprivation decreases somatic GAD65 puncta number on layer 2/3 pyramidal neurons in mouse visual cortex. Neural Plast 2009:415135.

Kumar S, Khisti R, Morrow A (2005) Regulation of native GABAA receptors by PKC and protein phosphatase activity. Psychopharmacology 183:241–247.

Kurotani T, Yamada K, Yoshimura Y, Crair M, Komatsu Y (2008) State-dependent bidirectional modification of somatic inhibition in neocortical pyramidal cells. Neuron 57:905–916.

Maffei A, Turrigiano G (2008a) Multiple modes of network homeostasis in visual cortical layer 2/3. J Neurosci 28:4377–4384.

Maffei A, Turrigiano G (2008b) The age of plasticity: developmental regulation of synaptic plasticity in neocortical microcircuits. Progress in Brain Research 169:211–223.

Maffei A, Fontanini A (2009) Network homeostasis: a matter of coordination. Current Opinion in Neurobiology 19:168–173.

Maffei A, Nataraj K, Nelson S, Turrigiano G (2006) Potentiation of cortical inhibition by visual deprivation. Nature 443:81–84.

Maffei A, Nelson S, Turrigiano G (2004) Selective reconfiguration of layer 4 visual cortical circuitry by visual deprivation. Nat Neurosci 7:1353–1359.

Mainardi M, Landi S, Berardi N, Maffei L, Pizzorusso T (2009) Reduced responsiveness to long-term monocular deprivation of parvalbumin neurons assessed by c-Fos staining in rat visual cortex. PloS One 4:e4342.

Marks C, Cheng K, Cummings D, Belluscio L (2006) Activity-dependent plasticity in the olfactory intrabulbar map. J Neurosci 26:11257–11266.

Marty S, Berzaghi Mda P, Berninger B (1997) Neurotrophins and activity-dependent plasticity of cortical interneurons. Trends Neurosci 20:198–202.

Miyoshi G, Butt SJ, Takebayashi H, Fishell G (2007) Physiologically distinct temporal cohorts of cortical interneurons arise from telencephalic Olig2-expressing precursors. J Neurosci 27:7786–7798.

Nothias F, Fishell G, Ruizi Altaba A (1998) Cooperation of intrinsic and extrinsic signals in the elaboration of regional identity in the posterior cerebral cortex. Curr Biol 8:459–462.

Nugent F, Penick E, Kauer J (2007) Opioids block long-term potentiation of inhibitory synapses. Nature 446:1086–1090.

Otsuka T, Kawaguchi Y (2009) Cortical inhibitory cell types differentially form intralaminar and interlaminar subnetworks with excitatory neurons. J Neurosci 29:10533–10540.

Rutherford L, Nelson S, Turrigiano G (1998) BDNF Has Opposite Effects on the Quantal Amplitude of Pyramidal Neuron and Interneuron Excitatory Synapses. Neuron 21:521–530.

Spatkowski G, Schilling K (2003) Postnatal dendritic morphogenesis of cerebellar basket and stellate cells in vitro. J Neurosci Res 72:317–326.

Sperry RW (1963) Chemoaffinity in the Orderly Growth of Nerve Fiber Patterns and Connections. Proc Natl Acad Sci USA 50:703–710.

Sugiyama S, Di Nardo AA, Aizawa S, Matsuo I, Volovitch M, Prochiantz A, Hensch TK (2008) Experience-dependent transfer of Otx2 homeoprotein into the visual cortex activates postnatal plasticity. Cell 134:508–520.

Sun Q (2009) Experience-dependent intrinsic plasticity in interneurons of barrel cortex layer IV. J Neurophysiol (in press).

Tagawa Y, Kanold P, Majdan M, Shatz C (2005) Multiple periods of functional ocular dominance plasticity in mouse visual cortex. Nat Neurosci 8:380–388.

Tan Z, Hu H, Huang Z, Agmon A (2008) Robust but delayed thalamocortical activation of dendritic-targeting inhibitory interneurons. Proc Natl Acad Sci 105:2187–2192.

Thomson A, Lamy C (2007) Functional maps of neocortical local circuitry. Front Neurosci 1:19–42.

Tyler W, Petzold G, Pal S, Murthy V (2007) Experience-dependent modification of primary sensory synapses in the mammalian olfactory bulb. J Neurosci 27:9427–9438.

Wang Y, Gupta A, Toledo-Rodriguez M, Wu CZ, Markram H (2002) Anatomical, physiological, molecular and circuit properties of nest basket cells in the developing somatosensory cortex. Cereb Cortex 12:395–410.

Wiesel TN, Hubel DH (1963a) Effects of Visual Deprivation on Morphology and Physiology of Cells in the Cats Lateral Geniculate Body. J Neurophysiol 26:978–993.

Wiesel TN, Hubel DH (1963b) Single-Cell Responses in Striate Cortex of Kittens Deprived of Vision in One Eye. J Neurophysiol 26:1003–1017.

Wiesel TN, Hubel DH (1965) Extent of recovery from the effects of visual deprivation in kittens. J Neurophysiol 28:1060–1072.

Wiesel TN, Hubel DH (1974) Ordered arrangement of orientation columns in monkeys lacking visual experience. J Comp Neurol 158:307–318.

Woodruff A, Yuste R (2008) Of Mice and men, and chandeliers. PloS Bio 6:e243.

Xu Q, Cobos I, De La Cruz E, Rubenstein JL, Anderson SA (2004) Origins of cortical interneuron subtypes. J Neurosci 24:2612–2622.

Yoon B, Smith G, Heynen A, Neve R, Bear M (2009) Essential role for a long-term depression mechanism in ocular dominance plasticity. Proc Natl Acad Sci USA 106:9860–9865.

Yu X, Sanes D, Aristizabal O, Wadghiri Y, Turnbull D (2007) Large-scale reorganization of the tonotopic map in mouse auditory midbrain revealed by MRI. Proc Natl Acad Sci 104: 12193–12198.

Chapter 2
Long-Term Modification at Inhibitory Synapses in Developing Visual Cortex

Yukio Komatsu and Yumiko Yoshimura

Abstract Involvement of bidirectional modification at excitatory synapses in experience-dependent cortical maturation has been supported by various experimental data in visual cortex. Experiments using slice preparations demonstrated that cortical inhibitory synapses also undergo long-term potentiation (LTP) and depression (LTD) during the critical period. High-frequency stimulation (HFS) of excitatory and inhibitory inputs to pyramidal neurons induces LTD at inhibitory synapses when it elicits depolarizing responses large enough to activate NMDA receptors. HFS induces inhibitory LTP instead when it fails to activate NMDA receptors. Thus, the direction of modification is determined by postsynaptic NMDA receptors. LTD induction requires Ca^{2+} entry via NMDA receptors, whereas LTP induction requires IP_3 receptor-mediated Ca^{2+} release, presumably triggered by $GABA_B$ receptor activation in the absence of substantial NMDA receptor activation. Intracellular Ca^{2+} release likely initiates BDNF release from the postsynaptic cell and activates TrkB receptors on inhibitory terminals, presumably leading to presynaptic enhancement of synaptic transmission. LTP maintenance requires presynaptic, but not postsynaptic, firing and associated Ca^{2+} entry at some intervals. This bidirectional modification at inhibitory synapses may contribute to the refinement and maintenance of visual responsiveness, and regulation of the critical period in visual cortex.

Introduction

Visual experience plays a crucial role in the maturation of visual cortical functions during the postnatal critical period (Wiesel 1982; Frégnac and Imbert 1984). It is generally assumed that activity-dependent long-term modification of cortical

Y. Komatsu (✉)
Department of Neuroscience, Research Institute of Environmental Medicine, Nagoya University, Nagoya 464-8601, Japan
e-mail: komatsu@riem.nagoya-u.ac.jp

M.A. Woodin and A. Maffei (eds.), *Inhibitory Synaptic Plasticity*,
DOI 10.1007/978-1-4419-6978-1_2,

synapses underlies this developmental process (Stent 1973; Bear et al. 1987; Singer 1995; Katz and Shatz 1996; Zhang and Poo 2001). Synaptic plasticity in visual cortex was examined using slice preparations and multiple types of long-term modification were demonstrated at excitatory synapses (Tsumoto 1992; Bear and Kirkwood 1993). The level of depolarization produced in postsynaptic neurons by conditioning stimulation is a crucial factor in the determination of the initiation and direction of plastic changes in both excitatory synaptic transmission (Malinow and Miller 1986; Gustafsson et al. 1987; Artola et al. 1990) and the visual responses of cortical neurons (Frégnac et al. 1988; Reiter and Stryker 1988). Initially, it was commonly considered that inhibitory synapses themselves did not undergo long-term modification physiologically, and that their role in synaptic plasticity was to regulate the modification of excitatory synaptic strength by modulating the postsynaptic depolarization resulting from excitatory inputs.

The visual responsiveness of cortical cells emerges from the spatiotemporal integration of excitatory and inhibitory synaptic inputs. Unless these two types of connections are properly organized in a balanced manner, visual stimulation may fail to produce an adequate level of visual responses, which are selective for stimulus features. This requirement may not be easily fulfilled, if only excitatory synapses are modifiable, suggesting that plastic changes can also occur at inhibitory synapses. In fact, our studies using slice preparations demonstrated that inhibitory synapses on pyramidal cells are highly modifiable during the critical period, as described below.

Involvement of NMDA Receptors in the Determination of the Direction of Modification in Synaptic Strength

In most of our experiments studying plasticity at γ-aminobutyric acid (GABA)-mediated inhibitory synapses, inhibitory postsynaptic potentials (IPSPs) or currents (IPSCs) evoked by stimulation of presynaptic fibers using bipolar metal electrodes placed in layer 4 were recorded from layer 5 pyramidal cells in rat visual cortical slices under a blockade of excitatory synaptic transmission using *N*-methyl-D-aspartate (NMDA) and non-NMDA receptor antagonists. These inhibitory responses evoked by test stimulation are mediated by $GABA_A$ receptors. Inhibitory synaptic transmission could undergo long-term potentiation (LTP) and depression (LTD) depending on the pharmacological manipulation when high-frequency stimulation (HFS) was applied to excitatory and inhibitory inputs (Komatsu and Iwakiri 1993). Under the blockade of non-NMDA and NMDA receptors, HFS eliciting hyperpolarizing responses in the postsynaptic cell induced LTP of IPSPs. When the same HFS was applied while the NMDA receptor antagonist 2-amino-5-phosphonovaleric acid (APV) was removed, it elicited hyperpolarizing or small depolarizing responses. After this stimulation, LTP also occurred at inhibitory synapses. However, when $GABA_A$ receptors were blocked by bicuculline methiodide together with the removal of APV, HFS elicited large depolarizing responses and LTD of IPSPs occurred instead. In an experiment conducted in a normal solution without any blockers, analysis of unitary IPSCs demonstrated that inhibitory synapses also underwent LTD, when HFS elicited large depolarizing responses in recorded cells (Yoshimura et al. 2003).

These observations suggest that the direction of change in the strength of inhibitory synapses depends on the level of postsynaptic depolarization during HFS (Fig. 2.1). When HFS produces postsynaptic depolarization large enough to activate NMDA receptors, it induces inhibitory LTD. On the other hand, when HFS fails to produce the depolarization necessary to activate NMDA receptors, it induces inhibitory LTP. Thus, NMDA receptors seem to determine the direction of change in inhibitory synaptic strength. This kind of bidirectional modification at GABAergic synapses was also demonstrated in neonate CA3 pyramidal neurons and developing tadpole tectal neurons (Mclean et al. 1996; Liu et al. 2007).

It is well known that NMDA receptors also play a crucial role in the determination of the direction of change in excitatory synaptic strength of CA1 pyramidal cells (Malenka and Nicoll 1993; Bear and Abraham 1996). The mechanism seems

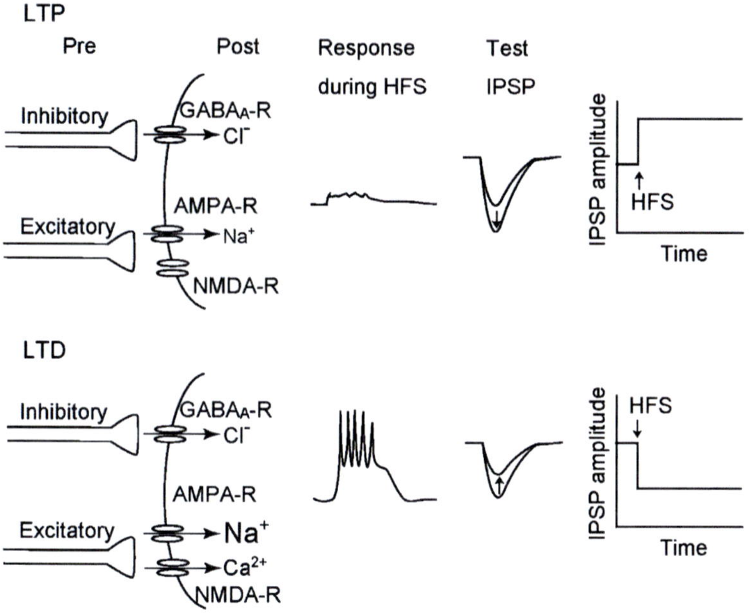

Fig. 2.1 Bidirectional modification at inhibitory synapses of pyramidal neurons in developing visual cortex

different between the excitatory and inhibitory synapses. At excitatory synapses, both LTP and LTD require NMDA receptor-mediated Ca^{2+} entry for induction. It is considered that a high level of Ca^{2+} increase results in LTP, while a moderate level of Ca^{2+} increase results in LTD (Malenka and Nicoll 1993; Lisman 1994). Inhibitory LTP requires no NMDA receptor activation in the postsynaptic cells.

Induction Mechanism of LTP

The induction mechanism of LTP at visual cortical inhibitory synapses examined so far is schematically illustrated in Fig. 2.2. LTP induction requires a postsynaptic elevation of Ca^{2+} like most LTP at excitatory synapses. While Ca^{2+} entry through NMDA receptors is necessary for the induction of inhibitory LTD (Komatsu and Iwakiri 1993; Yoshimura et al. 2003), intracellular Ca^{2+} release is responsible for the induction of inhibitory LTP (Komatsu 1996). The induction of LTP was blocked by postsynaptic loading of any of the following agents: G-protein inhibitor GDPβS, phospholipase C (PLC) inhibitor U73122, inositol trisphosphate (IP_3) receptor blocker heparin, and Ca^{2+} chelator BAPTA. This suggests that the following postsynaptic biochemical cascade is involved in LTP induction: activation of G protein coupled receptors, activation of PLC, IP_3 formation and IP_3 receptor-mediated Ca^{2+} release from the internal Ca^{2+} store.

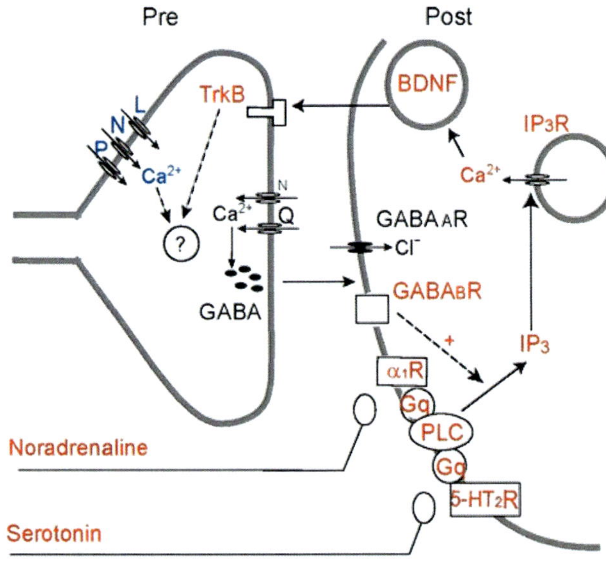

Fig. 2.2 Schematic drawing of molecular mechanisms for inhibitory LTP in visual cortical pyramidal neurons. Molecules indicated by *red* and *blue* colors are involved in the induction and maintenance of LTP, respectively

High-frequency activity of inhibitory synapses may trigger IP_3 receptor-mediated Ca^{2+} release, leading to LTP. LTP induction requires the activation of $GABA_B$ receptors, which are activated far more effectively by high-frequency than low-frequency stimulation, but not $GABA_A$ receptors (Komatsu 1996). $GABA_B$ receptors commonly couple to adenylate cyclase via Gi, but not PLC directly. However, it was reported that $GABA_B$ receptor activation facilitated α_1 adrenoceptor-mediated IP_3 formation in cerebral cortex (Crawford and Young 1990), suggesting that $GABA_B$ receptors can potentiate IP_3 receptor-mediated Ca^{2+} release via some indirect pathway. Activation of α_1 adrenoceptors and serotonin 5-HT_2 receptors, which couple to PLC, both facilitated LTP induction (Komatsu 1996). Thus, we consider, tentatively, that HFS-induced $GABA_B$ receptor activation facilitates α_1 adrenoceptor- and/or 5-HT_2 receptor-mediated IP_3 formation and initiates IP_3 receptor-mediated Ca^{2+} release (Fig. 2.2). However, additional investigation is required to substantiate this supposition.

Activity-Dependent Maintenance of LTP

A transient intracellular Ca^{2+} elevation can initiate long-term modification of synaptic strength. In NMDA receptor-dependent LTP at excitatory synapses, it has been proposed that a temporary Ca^{2+} elevation mediated by NMDA receptor activation produces autophosphorylation of CaM kinase II, converting the enzyme into a Ca^{2+}-independent active form, which phosphorylates its substrates persistently (Lisman 1994). This biochemical process may maintain potentiation long-lastingly without any following neural activity. Inhibitory LTP in visual cortex requires neural activity for maintenance, although the necessary frequency of activity is low (Komatsu and Yoshimura 2000). Although LTP persisted as long as test stimulation was continued at 0.1 Hz, it disappeared often after temporal cessation of test stimulation (30 min) and always after a temporary pharmacological blockade of action potentials. When postsynaptic responses were completely blocked by GABA receptor antagonists instead of sodium channel blockers, LTP persisted, indicating that LTP maintenance requires presynaptic, but not postsynaptic, neural activity.

Multiple types of high-threshold voltage-gated Ca^{2+} channels located at a presynaptic site are likely involved in maintenance (Komatsu and Yoshimura 2000), as shown in Fig. 2.2. Transmitter release at these inhibitory synapses was mediated mostly by Q-type Ca^{2+} channels, and partially by N-type Ca^{2+} channels. LTP maintenance was completely blocked by a temporary pharmacological blockade of either P or L-type Ca^{2+} channels, which are not involved in triggering transmitter release at these synapses. In addition, a brief bath application of membrane permeable slow Ca^{2+} chelator EGTA-AM abolished LTP without affecting basal synaptic transmission. These results suggest that LTP maintenance requires a slow presynaptic Ca^{2+} elevation at some intervals, which activates Ca^{2+}-dependent reactions different from that mediating transmitter release.

Expression Site of LTP

The presence of a maintenance mechanism at the presynaptic site may be in favor of presynaptic expression of LTP because of simplicity of the mechanism. To test this possibility, we analyzed the coefficient of variation (CV) and paired-pulse ratio (PPR) of IPSC amplitude (Inagaki et al. 2008). CV decreased in association with LTP, consistent with a presynaptic locus of LTP expression (Faber and Korn 1991; Manabe et al. 1993). However, PPR remained almost unchanged. These results are not inconsistent with the supposition that this inhibitory LTP is expressed, at least in part, presynaptically by the recruitment of more release sites. However, we consider that this issue must be examined more rigorously.

Retrograde Signaling During LTP Induction

If LTP is indeed expressed presynaptically as discussed above, some information must be sent backwards from the postsynaptic to the presynaptic cells during induction. We considered that brain-derived neurotrophic factor (BDNF) was a promising candidate for mediating the signal, based on the following observations. BDNF is expressed in pyramidal neurons but not inhibitory interneurons in sensory cortex, while TrkB is expressed in both types of neurons (Cellerino et al. 1996; Rocamora et al. 1996; Gorba and Wahle 1999). BDNF can be released from the somatodendritic domain of pyramidal neurons in an activity-dependent manner through intracellular Ca^{2+} elevation, and IP_3 receptor-mediated Ca^{2+} increase can initiate BDNF release (Lessmann et al. 2003). It was reported that BDNF immunoreactivity associated with vesicular-like structures was detected in both cell body and processes of layer 5 neurons of rat cortex (Fawcett et al. 1997). Furthermore, it was demonstrated using cultured cortical neurons that BDNF released from pyramidal neurons facilitated the establishment of GABAergic synaptic terminals to those neurons and increased the frequency, but not the amplitude, of miniature IPSCs (Kohara et al. 2007), which is consistent with the recruitment of new release sites.

Thus, we tested the possibility that BDNF mediates a retrograde signal for LTP induction (Inagaki et al. 2008). LTP production was prevented by bath application of TrkB-IgG which blocks the action of BDNF on TrkB (Shelton et al. 1995) or function-blocking anti-BDNF antibody. In addition, LTP did not occur when HFS was applied in the presence of K252a, which is membrane permeable and likely inhibits both pre and postsynaptic TrkB receptor tyrosine kinases (Knüsel and Hefti 1992), whereas LTP did occur when bath application of K252a was started soon after HFS. Postsynaptic loading of K252a, which may inhibit postsynaptic but not presynaptic TrkB receptors, did not affect LTP production. Thus, it is likely that LTP induction requires the activation of TrkB receptors at the presynaptic site by BDNF released from postsynaptic cells. In BAPTA-loaded cells, which are incapable of Ca^{2+}-dependent BDNF release, HFS failed to induce LTP and this failure was rescued by bath application of BDNF during HFS. These lines of evidence strongly suggest that HFS initiates IP_3 receptor-mediated Ca^{2+} release, leading to BDNF release from the somatodendritic domain of pyramidal neurons (Fig. 2.2). The released BDNF may then activate TrkB at the presynaptic site, leading to the enhancement of synaptic strength. In BAPTA-loaded cells, LTP occurred only at the synapses activated by HFS applied during bath application of BDNF. Therefore, it is likely that HFS has also to activate some process at the presynaptic site, other than TrkB activation, in order to induce LTP. This additional process may assure the input specificity of this LTP.

BDNF plays a similar role in GABAergic LTP in CA3 pyramidal neurons and tadpole optic tectum neurons, although the postsynaptic Ca^{2+} increase responsible for BDNF release may be ascribed to voltage-gated Ca^{2+} channels and NMDA receptors for the former and latter cells, respectively (Gubellini et al. 2005; Liu et al. 2007). Thus, it is suggested that BDNF may commonly act as a retrograde messenger in LTP at GABAergic synapses during development.

Possible Roles of Inhibitory Synaptic Plasticity in the Refinement of Visual Function

It is generally considered that long-term modification at excitatory synapses is crucial for activity-dependent refinement of visual responsiveness. There is some experimental evidence supporting this hypothesis (Heynen et al. 2003; Yoshimura et al. 2008). It is likely that inhibitory plasticity also contributes to the maturation of visual function. Inhibitory LTP occurred mostly during development and its production is strongly influenced by noradrenaline and serotonin (Komatsu 1994, 1996), as was the case for visual response plasticity of cortical cells (Kasamatsu and Pettigrew 1976; Gu and Singer 1995). The maintenance of visual responsiveness shaped through visual experience requires visual inputs for the following time period during the critical period, as the maintenance of inhibitory LTP requires inputs to potentiated synapses. For example, orientation selectivity is improved to a level almost the same as that for adults in the middle of the critical period (Frégnac and Imbert 1984) and a following deprivation of light, even for a few days, degrades the selectivity (Freeman et al. 1981). These lines of evidence support the hypothesis that inhibitory LTP contributes to the experience-dependent refinement and maintenance of visual responsiveness during development.

Inhibitory synaptic connections can enhance selective responsiveness established by excitatory connections (Sillito 1975; Sompolinsky and Shapley 1997). Repetitive visual stimulation, evoking depolarizing responses large enough to activate NMDA receptors of a pyramidal cell, can induce LTD at inhibitory synapses on the cell. On the contrary, repetitive visual stimulation, evoking postsynaptic responses smaller than those necessary to activate NMDA receptors, can induce inhibitory LTP. If visual stimulation induces inhibitory LTP in a cortical cell, the stimulation evokes smaller visual responses in the cell thereafter. However, if visual stimulation induces inhibitory LTD instead, the stimulation evokes larger visual responses thereafter. Thus, small visual responses get smaller, whereas large visual responses get larger, leading to the enhancement of response selectivity. Recently, it was demonstrated that inhibitory LTP contributed to the decreased visual response in cortical neurons following visual deprivation (Maffei et al. 2006).

Interestingly, inhibitory plasticity in tadpole tectal neurons plays a functional role, which is in remarkable contrast to that in visual cortex (Tao and Poo 2005; Liu et al. 2007). In these cells, inhibitory synapses underwent LTP when visual stimulation produced depolarizing responses large enough to activate postsynaptic NMDA receptors, but they underwent LTD when the stimulation failed to produce such depolarization. Thus, this modification can adjust the strength of inhibitory synapses in proportion to that of excitatory synapses converging to the same cell. This process may contribute to the establishment of overlapped receptive fields for excitatory and inhibitory visual inputs and an adequate level of responsiveness to visual stimulation.

Inhibitory plasticity may be involved in the regulation of the critical period in addition to the modification of visual responsiveness. The start of critical period is regulated by the maturation of cortical inhibition (Hensch 2005). Dark rearing from birth retards the maturation of inhibitory synaptic transmission (Morales et al. 2002) and the start of the critical period (Mower 1991). BDNF is required for morphological and functional maturation of neocortical inhibitory neurons (Rutherford et al. 1997; Jin et al. 2003; Kohara et al. 2003, 2007; Itami et al. 2007). In mice in which BDNF was overexpressed, maturation of cortical inhibition and the start of the critical period both proceeded earlier than in normal mice (Huang et al. 1999). The production of BDNF is experience dependent, and the expression level is very low before eye opening and increases steeply after eye opening in visual cortex (Castrén et al. 1992; Schoups et al. 1995; Katoh-Semba et al. 1997; Pollock et al. 2001). Thus, it is likely that visual inputs increase the amount of BDNF expressed in pyramidal neurons and a resulting increase in BDNF release facilitates the maturation of inhibitory synaptic inputs to those neurons. Because BDNF likely mediates a retrograde signal for the induction of LTP at inhibitory synapses, we consider that inhibitory LTP can contribute to the regulation of the critical period.

References

Artola A, Bröcher S, Singer W (1990) Different voltage-dependent thresholds for inducing long-term depression and long-term potentiation in slices of rat visual cortex. Nature 347: 69–72

Bear FM, Abraham WC (1996) Long-term depression in hippocampus. Annu Rev Neurosci 19: 437–462

Bear MF, Cooper LN, Ebner FF (1987) A physiological basis for a theory of synapse modification. Science 237: 42–48

Bear MF, Kirkwood A (1993) Neocortical long-term potentiation, Curr Opin Neurobiol 3: 197–202

Castrén E, Zafra F, Thoenen H et al. (1992) Light regulates expression of brain-derived neurotrophic factor mRNA in rat visual cortex. Proc Natl Acad Sci USA 89: 9444–9448

Cellerino A, Maffei L, Domenici L (1996) The distribution of brain-derived neurotrophic factor and its receptor trkB in parvalbumin-containing neurons of the rat visual cortex. Eur J Neurosci 8: 1190–1197

Crawford MLA, Young JM (1990) Potentiation by γ-aminobutyric acid of α_1-agonist-induced accumulation of inositol phosphates in slices of rat cerebral cortex. J Neurochem 54: 2100–2109

Faber DS, Korn H (1991) Applicability of the coefficient of variation method for analyzing synaptic plasticity. Biophys J 60: 1288–1294

Fawcett JP, Aloyz R, McLean JH et al. (1997) Detection of brain-derived neurotrophic factor in a vesicular fraction of brain synaptosomes. J Biol Chem 272: 8837–8840

Freeman RD, Mallach R, Hartley S (1981) Responsivity of normal kitten striate cortex deteriorates after brief binocular deprivation. J Neurophysiol 45: 1074–1084

Frégnac Y, Imbert M (1984) Development of neuronal selectivity in primary visual cortex of cat, Physiol Rev 64: 325–434

Frégnac Y, Shulz D, Thorpe S et al. (1988) A cellular analogue of visual cortical plasticity. Nature 333: 367–370

Gorba T, Wahle P (1999) Expression of trkB and trkC but BDNF mRNA in neurochemically identified interneurons in rat visual cortex *in vivo* and in organotypic cultures. Eur J Neurosci 11: 1179–1190

Gu Q, Singer W (1995) Involvement of serotonin in developmental plasticity of kitten visual cortex. Eur J Neurosci 7: 1146–1153

Gubellini P, Ben-Ari Y, Gaïarsa J-L (2005) Endogenous neurotrophins are required for the induction of GABAergic long-term potentiation in the neonatal rat hippocampus. J Neurosci 25: 5796–5802

Gustafsson B, Wigström H, Abraham WC et al. (1987) Long-term potentiation in the hippocampus using depolarizing current pulses as the conditioning stimulus to single volley synaptic potentials. J Neurosci 7: 774–780

Hensch TK (2005) Critical period plasticity in local cortical circuits. Nature Rev Neurosci 6: 877–888

Heynen AJ, Yoon B-J, Liu C-H et al. (2003) Molecular mechanism for loss of visual cortical responsiveness following brief monocular deprivation. Nat Neurosci 6: 854–862

Huang ZJ, Kirkwood A, Pizzorusso T et al. (1999) BDNF regulates the maturation of inhibition and the critical period of plasticity in mouse visual cortex. Cell 98: 739–755

Inagaki T, Begum T, Reza F et al. (2008) Brain-derived neurotrophic factor-mediated retrograde signaling required for the induction of long-term potentiation at inhibitory synapses of visual cortical pyramidal neurons. Neurosci Res 61: 192–200

Itami C, Kimura F, Nakamura S (2007) Brain-derived neurotrophic factor regulates the maturation of layer 4 fast-spiking cells after the second postnatal week in the developing barrel cortex. J Neurosci 27: 2241–2252

Jin X, Hu H, Mathers PH et al. (2003) Brain-derived neurotrophic factor mediates activity-dependent dendritic growth in nonpyramidal neocortical interneurons in developing organotypic cultures. J Neurosci 23: 5662–5673

Kasamatsu T, Pettigrew JD (1976) Depletion of brain catecholamine: failure of ocular dominance shift after monocular occlusion in kitten. Science 194: 206–209

Katoh-Semba R, Takeuchi IK, Semba R et al. (1997) Distribution of brain-derived neurotrophic factor in rats and its changes with development in the brain. J Neurochem 69: 34–42

Katz LC, Shatz CJ (1996) Synaptic activity and the construction of cortical circuits. Science 274: 1133–1138

Knüsel B, Hefti F (1992) K-252 compounds: modulators of neurotrophin signal transduction. J Neurochem 59: 1987–1996

Kohara K, Kitamura A, Adachi N et al. (2003) Inhibitory but not excitatory cortical neurons require presynaptic brain-derived neurotrophic factor for dendritic development, as revealed by chimera cell culture. J Neurosci 23: 6123–6131

Kohara K, Yasuda H, Huang Y et al. (2007). A local reduction in cortical GABAergic synapses after a loss of endogenous brain-derived neurotrophic factor, as revealed by single-cell gene knock-out method. J Neurosci 27: 7234–7244
Komatsu Y (1994) Age-dependent long-term potentiation of inhibitory synaptic transmission in rat visual cortex. J Neurosci 14: 6488–6499
Komatsu Y (1996) $GABA_B$ receptors, monoamine receptors, and postsynaptic inositol trisphosphate-induced Ca^{2+} release are involved in the induction of long-term potentiation at visual cortical inhibitory synapses. J Neurosci 16: 6342–6352
Komatsu Y, Iwakiri M (1993) Long-term modification of inhibitory synaptic transmission in developing visual cortex. Neuroreport 4: 907–910
Komatsu Y, Yoshimura Y (2000) Activity-dependent maintenance of long-term potentiation at visual cortical inhibitory synapses. J Neurosci 20: 7539–7546
Lessmann V, Gottmann K, Malcangio M (2003) Neurotrophin secretion: current facts and future prospects. Prog Neurobiol 69: 341–374
Lisman J (1994) The CaM kinase II hypothesis for the storage of synaptic memory. Trends Neurosci 17: 406–7546
Liu Y, Zhang LI, Tao HW (2007) Heterosynaptic scaling of developing GABAergic synapses: dependence on glutamatergic input and developmental stage. J Neurosci 27: 5301–5312
Maffei A, Nataraj K, Nelson SB et al. (2006) Potentiation of cortical inhibition by visual deprivation. Nature 443: 81–84
Malenka RC, Nicoll RA (1993) NMDA-receptor-dependent synaptic plasticity: multiple forms and mechanisms. Trends Neurosci 16: 521–527
Malinow R, Miller JP (1986) Postsynaptic hyperpolarization during conditioning reversibly blocks induction of long-term potentiation. Nature 320: 529–530
Manabe T, Wyllie DJA, Perkel DJ et al. (1993) Modulation of synaptic transmission and long-term potentiation: effects on paired pulse facilitation and EPSC variance in the CA1 region of the hippocampus. J Neuophysiol 70: 1451–1459
McLean HA, Caillard O, Ben-Ari Y et al. (1996) Bidirectional plasticity expressed by GABAergic synapses in the neonatal rat hippocampus. J Physiol Lond 496: 471–477
Morales B, Choi S-Y, Kirkwood A (2002) Dark rearing alters the development of GABAergic transmission in visual cortex. J Neurosci 22: 8084–8090
Mower GD (1991) The effect of dark rearing on the time course of the critical period in cat visual cortex. Dev Brain Res 58: 151–158
Pollock GS, Vernon E, Forbes ME et al. (2001) Effects of early visual experience and diurnal rhythms on BDNF mRNA and protein levels in the visual system, hippocampus, and cerebellum. J Neurosci 21: 3923–3931
Reiter HO, Stryker MP (1988) Neural plasticity without postsynaptic action potentials: less-active inputs become dominant when kitten visual cortical cells are pharmacologically inhibited. Proc Natl Acd Sci USA 85: 3623–3627
Rocamora N, Welker E, Pascual M et al. (1996) Upregulation of BDNF mRNA expression in the barrel cortex of adult mice after sensory stimulation. J Neurosci 16: 4411–4419
Rutherford LC, DeWan A, Lauer HM et al. (1997) Brain-derived neurotrophic factor mediates the activity-dependent regulation of inhibition in neocortical cultures. J Neurosci 17: 4527–4535
Schoups AA, Elliott RC, Friedman WJ et al. (1995) NGF and BDNF are differentially modulated by visual experience in the developing geniculocortical pathway. Dev Brain Res 86: 326–334
Shelton DL, Sutherland J, Gripp J et al. (1995) Human trks: molecular cloning, tissue distribution, and expression of extracellular domain immunoadhesins. J Neurosci 15: 477–491
Sillito AM (1975) The contribution of inhibitory mechanisms to the receptive field properties of neurones in the striate cortex of the cat. J Physiol Lond 250: 305–1975
Singer W (1995) Development and plasticity of cortical processing architectures. Science 270: 758–764
Sompolinsky H, Shapley R (1997) New perspectives on the mechanisms for orientation selectivity. Curr Opin Neurobiol 7: 514–522

Stent G (1973) A physiological mechanism for Hebb's postulate of learning. Proc Nat Acd Sci USA 70: 997–1001

Tao HW, Poo M-M (2005) Activity-dependent matching of excitatory and inhibitory inputs during refinement of visual receptive fields. Neuron 45: 829–836

Tsumoto T (1992) Long-term potentiation and long-term depression in the neocortex. Prog Neurobiol 39: 209–228

Wiesel TN (1982) Postnatal development of the visual cortex and the influence of environment. Nature 299: 583–592

Yoshimura Y, Inaba M, Yamada K et al. (2008) Involvement of T-type Ca^{2+} channels in the potentiation of synaptic and visual responses during the critical period in rat visual cortex. Eur J Neurosci 28: 730–743

Yoshimura Y, Ohmura T, Komatsu Y (2003) Two forms of synaptic plasticity with distinct dependence on age, experience and NMDA receptor subtype in rat visual cortex. J Neurosci 23: 6557–6566

Zhang LI, Poo M-M (2001) Electrical activity and development of neural circuits. Nat Neurosci Suppl 4: 1207–1214

Chapter 3
Molecular Mechanism of Long-Term Plasticity at Cerebellar Inhibitory Synapses

Shin-ya Kawaguchi and Tomoo Hirano

Abstract The efficacy of synaptic transmission changes depending on the neuronal activity in the central nervous system. Such synaptic plasticity underlies experience-dependent refinement of information processing in a neuronal network, and is regarded as a cellular basis for learning and memory. Compared with excitatory synapses, little has been clarified about the regulatory mechanism of plasticity at inhibitory synapses. In this chapter, we summarize recent advances in understanding the molecular mechanism of inhibitory synaptic plasticity in the cerebellum. The GABAergic synapses on a Purkinje neuron undergo long-term potentiation of postsynaptic $GABA_A$ receptor ($GABA_AR$) responsiveness in response to the postsynaptic depolarization, which is called rebound potentiation (RP). The mechanism of RP regulation has been studied at the molecular level using electrophysiological experiments combined with molecular biological techniques, fluorescent imaging and systems biological computer simulation. We here describe how the induction of RP is regulated through complicated interaction of intracellular signaling cascades including protein kinases/phosphatases, and how a $GABA_AR$ binding protein is implicated in the establishment and maintenance of inhibitory synaptic plasticity.

Plasticity at Inhibitory Synapses

Information processing in a neuronal network is refined by experience through the neuronal activity-dependent synaptic plasticity, which has been regarded as a cellular basis for learning and memory (Malenka and Nicoll 1999; Bailey et al. 2000; Ito 2001; Lisman and Zhabotinsky 2001; Hansel et al. 2001; Kandel 2001). Synaptic plasticity has been studied at both excitatory and inhibitory synapses, although relatively limited information is available for inhibitory synaptic plasticity.

T. Hirano (✉)
Department of Biophysics, Graduate School of Science, Kyoto University, Sakyo-ku, Kyoto, Japan 606-8502
e-mail: thirano@neurosci.biophys.kyoto-u.ac.jp

M.A. Woodin and A. Maffei (eds.), *Inhibitory Synaptic Plasticity*,
DOI 10.1007/978-1-4419-6978-1_3, © Springer Science+Business Media, LLC 2011

Long-term potentiation (LTP) and depression (LTD) are induced at inhibitory synapses in various regions of the central nervous system such as the hippocampus, cerebral cortex, cerebellum, deep cerebellar nucleus, brain stem, and lateral superior olive (Komatsu 1994, 1996; Kano 1995; Marty and Llano 1995; Nusser et al. 1998; Aizenman et al. 1998; Gaiarsa et al. 2002).

In the cerebellum, inhibitory synapses on a Purkinje neuron show four types of synaptic plasticity: depolarization-induced suppression of inhibition (DSI), depolarization-induced potentiation of inhibition (DPI), rebound potentiation (RP), and LTD (Llano et al. 1991; Kano et al. 1992; Duguid and Smart 2004; Mittmann and Hausser 2007). They seem to contribute to the refined function of cerebellar neuronal network in concert with excitatory synaptic activities and plasticity. All of the above synaptic plasticity are triggered by activation of a climbing fiber that causes potent depolarization and the resultant increase in intracellular Ca^{2+} concentration in a Purkinje neuron. DSI and DPI are short-term plasticity accompanied by the altered probability of presynaptic GABA release, and each induction is mediated by a retrograde messenger molecule released from a postsynaptic Purkinje neuron (Llano et al. 1991, Duguid and Smart 2004). RP is expressed as the sustained alteration in properties and/or numbers of postsynaptic $GABA_AR$ (Kano et al. 1992; Kawaguchi and Hirano 2000). Finally LTD is induced when the presynaptic inhibitory interneuron and the climbing fiber are stimulated simultaneously at 1 Hz for 5 min (Mittmann and Hausser 2007), although the mechanism of LTD induction remains unknown. In this chapter, we focus on the regulation of RP as a model of postsynaptic plasticity at inhibitory synapses, and summarize recent advances in understanding the molecular mechanism.

Regulation of RP

RP is induced by the postsynaptic depolarization such as that caused by repetitive activation of a climbing fiber, and lasts for longer than tens of minutes (Kano et al. 1992). RP is expressed as the increased responsiveness of postsynaptic $GABA_AR$, and the RP induction depends on the depolarization-induced increase in intracellular Ca^{2+} concentration (Fig. 3.1). Activation of Ca^{2+}/calmodulin (Ca^{2+}/CaM)-dependent protein kinase II (CaMKII) downstream of the Ca^{2+} increase is required for the induction of RP (Kano et al. 1996). CaMKII phosphorylates $GABA_AR$ β and γ subunits in the cytoplasmic loop between the third and fourth transmembrane segments and augments the $GABA_AR$ responsiveness to GABA (Moss and Smart 1996; Brandon et al. 2002; Houston and Smart 2006). Thus, the direct phosphorylation of $GABA_AR$ by CaMKII might be involved in RP.

RP induction is regulated by an intriguing mechanism. Presynaptic activation of an inhibitory interneuron coupled with the postsynaptic depolarization suppresses the RP induction (Kawaguchi and Hirano 2000) (Fig. 3.1). This suppressive effect of presynaptic activation on the RP induction is mediated by activation of postsynaptic $GABA_B$ receptor ($GABA_BR$). Signaling pathways underlying the

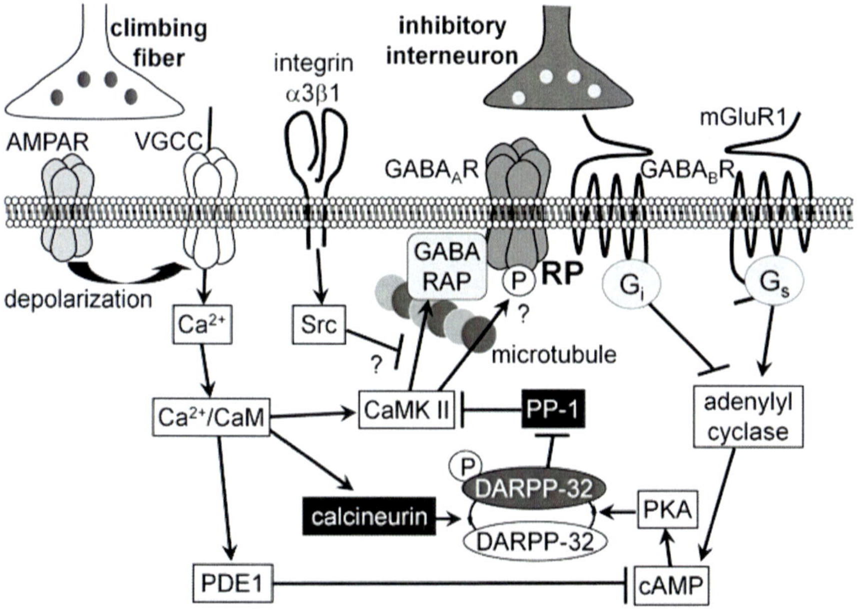

Fig. 3.1 Signaling cascades regulating the induction of RP. Depolarization caused by synaptic inputs from a climbing fiber causes large Ca^{2+} influx through voltage-gated Ca^{2+} channel. Ca^{2+}/CaM activates CaMKII which phosphorylates GABA$_A$R and also leads to the sustained conformation change of GABARAP, which would augment the function of individual GABA$_A$R though association with GABA$_A$R γ2 subunit and with microtubule. GABA$_B$R activation decreases the PKA activity through inhibition of adenylyl cyclase, and releases PP-1 from the inhibition by phospho-DARPP-32 in cooperation with calcineurin activated by the increase in intracellular Ca^{2+}. As a result, the CaMKII activity is inhibited by PP-1, and the RP induction is suppressed. Ca^{2+}/CaM-dependent PDE1 indirectly suppresses the CaMKII activity in a feed-forward manner through the hydrolysis of cAMP and the resultant reduction of PKA activity. A cell adhesion molecule integrin α3β1 negatively regulates the induction of RP through Src-family protein tyrosine kinase

GABA$_B$R-mediated suppression of RP have been clarified (Kawaguchi and Hirano 2000, 2002), as explained in detail in the following section.

The negative regulation of the induction of synaptic plasticity by presynaptic activity is unique, as most forms of synaptic plasticity reported to date are positively regulated by presynaptic activity. RP is induced not by homosynaptic activation but by the heterosynaptic excitatory inputs such as those from a climbing fiber (Kano et al. 1992). Taking it into consideration that repetitive activation of a climbing fiber causes the large Ca^{2+} increase in wide dendritic regions in a Purkinje neuron, the transmission efficacy at most of inhibitory synapses on the Purkinje neuron would be augmented at the same time without synapse-specificity. On the other hand, the suppressive effect of GABA$_B$R activation on the RP induction is confined to the synapses where postsynaptic GABA$_B$R is activated (Kawaguchi and Hirano 2000). Thus, the RP suppression by presynaptic activity would enable synapse-specific regulation of RP.

In addition to neuronal activity, a signal arising from a cell-adhesion molecule is also implicated in the regulation of RP (Kawaguchi and Hirano 2006). When activity of integrin, a kind of cell-adhesion molecule working as a receptor for extracellular matrix proteins, is facilitated, the RP induction is impaired, as it is abolished by a function blocking antibody against integrin α3 subunit or that against β1 subunit. These results suggest that the activity of integrin α3β1 heterodimer negatively regulates the induction of RP (Fig. 3.1). Integrin α3β1 seems to suppress RP through Src-family protein tyrosine kinase, although the detailed mechanism remains unclarified. Thus, the transmission efficacy of inhibitory synapses on a Purkinje neuron is elaborately regulated by neuronal activity and cell-extracellular matrix interactions.

Molecular Network that Controls the Induction and Suppression of RP

Regulation of protein kinases and phosphatases plays a critical role in the switching between the induction and the suppression of RP in response to different combinations of pre- and postsynaptic activation (Kawaguchi and Hirano 2002). In addition to CaMKII, the activity of cAMP-dependent protein kinase (PKA) is required for the induction of RP (Fig. 3.1). PKA contributes to the RP induction through gating the CaMKII activation by modulating the activity of counteracting protein phosphatase 1 (PP-1). When activated by Ca^{2+}/CaM, CaMKII autophosphorylates Thr286 residue, which turns the CaMKII into a Ca^{2+}-independent active form (Miller and Kennedy 1986; Giese et al. 1998). The phospho-Thr286 of CaMKII is dephosphorylated by PP-1. Further, PP-1 dephosphorylates proteins phosphorylated by CaMKII. Thus, the effective CaMKII activity needs adequate suppression of the PP-1 activity. PKA phosphorylates DARPP-32 (dopamine and cAMP-regulated phospho-protein 32 kDa), and the phosphorylated DARPP-32 negatively regulates the PP-1 activity through direct binding (Hemmings et al. 1984). Therefore, PKA supports the CaMKII activity by suppressing the PP-1 activity through phosphorylation of DARPP-32. On the other hand, the phosphorylated DARPP-32 is dephosphorylated by calcineurin (also called PP2B), which is activated by Ca^{2+} and calmodulin.

$GABA_B$R reduces the intracellular concentration of cAMP through inhibition of adenylyl cyclase via Gi/o-type trimeric G protein, decreasing PKA activity (Fig. 3.1). The basal level of cAMP is likely to be maintained by the activity of receptors coupled to Gs including metabotropic glutamate receptor mGluR1 (Sugiyama et al. 2008). The suppression of RP induction by $GABA_B$R depends on the activities of calcineurin, DARPP-32, and PP-1 (Kawaguchi and Hirano 2002). It is explained that when the PKA activity is reduced together with the calcineurin activation in response to coincident depolarization and $GABA_B$R activation, DARPP-32 is effectively dephosphorylated, resulting in PP-1 release from inhibition by phospho-DARPP-32. As a result, CaMKII activity is inhibited by augmented PP-1 activity, impairing the RP induction (Fig. 3.1).

Systems Biological Study on Signaling Cascades Regulating RP

The overall framework of RP regulation shown in Fig. 3.1 was presumed by combining pieces of electrophysiological experimental data. Therefore, it is unclear whether the signaling cascades actually operate as supposed during the induction or suppression of RP. To address this issue, a kinetic simulation model of the signaling cascades regulating RP induction was recently constructed. Each biochemical reaction was expressed with a differential equation, and the behavior of the molecular network was systematically analyzed by numerically solving the differential equations (Kitagawa et al. 2009). Systems biological modeling of molecular networks is useful for quantitative understanding of the complicated and supra-linear cellular behaviors such as oscillation, hysteresis, adaptation, bistability, etc (Bhalla and Iyengar 1999; Brandman and Meyer 2008).

Simulation of the model reproduces the RP induction or suppression as the sustained CaMKII activation or its suppression in response to the conditioning transient Ca^{2+} increase alone or coupled with $GABA_BR$ activation, respectively. Thus, the scheme of signaling cascades mediating the induction or suppression of RP is confirmed. The sustained CaMKII activation in response to the transient Ca^{2+} increase seems to be a basis of long-lasting maintenance of RP. The function of $GABA_AR$ composed of $\alpha 1$, $\beta 2$, and $\gamma 2$ subunits is positively regulated by CaMKII-mediated phosphorylation (Houston et al. 2008). Thus, the sustained phosphorylation of $GABA_AR$ by long-term CaMKII activation might play a role in the expression and maintenance of RP, together with a kind of $GABA_AR$-binding protein, as described in the next section.

In addition to the confirmation of switching mechanism of RP, the model analyses provide two insights on the molecular mechanism of RP regulation (Kitagawa et al. 2009). One is that the sustained CaMKII activation upon the transient Ca^{2+} increase is robustly brought about by the synergistic actions of two positive feedback pathways: the CaMKII autophosphorylation at Thr286 and the inhibition of PDE1 (a type of Ca^{2+}/CaM-dependent phosphodiesterase) by CaMKII phosphorylation (Fig. 3.1). As noted above, CaMKII autophosphorylation at Thr286 renders the enzyme Ca^{2+}/CaM-independently active. Phosphorylation of PDE1 by CaMKII decreases the affinity of PDE1 for Ca^{2+}/CaM (Hashimoto et al. 1989). The decreased affinity of PDE1 for Ca^{2+}/CaM slows down the hydrolysis of cAMP, resulting in augmentation of the PKA activity. As a result, phosphorylation level of DARPP-32 is increased, facilitating inhibition of PP-1. Thus, CaMKII facilitates its own activity through two positive feedback pathways, autophosphorylation and indirect down-regulation of the counteracting PP-1 activity through PDE1 inhibition. The simulation shows that weakening either the CaMKII autophosphorylation or the PDE1 inhibition by CaMKII impairs the sustained CaMKII activation.

The other new information provided by analyses on the model is about a mechanism to control the Ca^{2+} threshold for RP induction (Kitagawa et al. 2009). The molecular network regulating RP contains two feed-forward inhibitory pathways to the CaMKII activity (Fig. 3.1): one is a pathway through calcineurin, and the other is a pathway through PDE1. Both calcineurin and PDE1 are activated by Ca^{2+}/CaM, and release PP-1 from inhibition by phospho-DARPP-32, suppressing the CaMKII activity. Thus, these pathways are likely involved in the control of the Ca^{2+} threshold for RP induction. Model simulation shows that the Ca^{2+} threshold for RP induction is markedly lowered by the decrease in PDE1 activity, but not in calcineurin activity (Fig. 3.2), suggesting that the Ca^{2+} threshold for RP induction is predominantly regulated by the feed-forward pathway through PDE1. This prediction was examined by electrophysiological experiments on cultured Purkinje neurons (Kitagawa et al. 2009). In the presence of inhibitors for PDE1, but not for calcineurin, RP is successfully induced by a weaker conditioning depolarization which causes less Ca^{2+} increase, validating the model prediction (Fig. 3.2). Thus, combined application of model simulation and electrophysiological experiments revealed that PDE1 predominantly determines the Ca^{2+} threshold for RP induction.

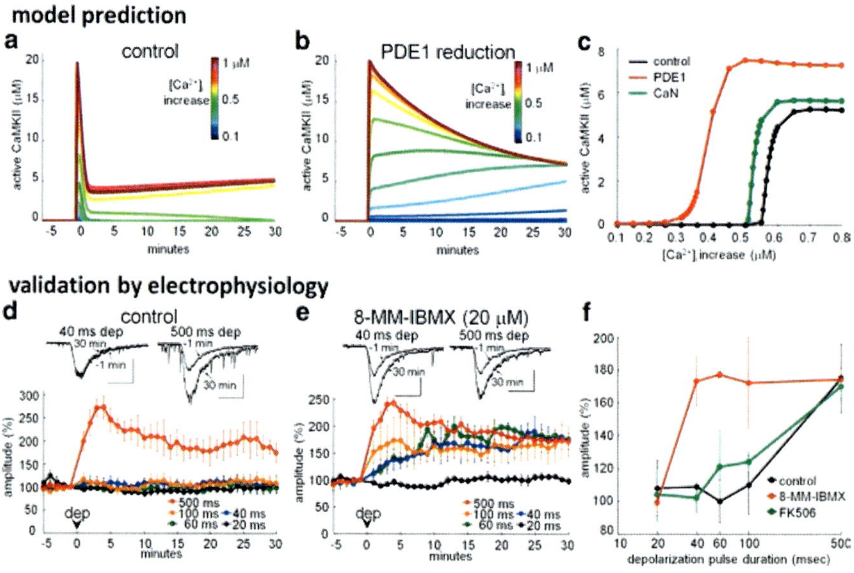

Fig. 3.2 Model prediction (**a–c**) and experimental validation (**d–f**) of a critical role of PDE1 in regulation of the Ca^{2+} threshold for RP induction. (**a, b**) Time courses of the simulated active CaMKII amount with (**b**) or without (**a**) 80% reduction of PDE1 in response to the transient Ca^{2+} increase (10 s) with various amplitudes. (**c**) Relation of the amplitude of transient Ca^{2+} increase and the CaMKII activity at 30 min without or with 80% reduction of PDE1 or calcineurin. (**d, e**), Representative current traces and time courses of amplitudes of GABA responses before and after the conditioning stimulation consisting of different durations of depolarization pulses in the absence (**d**) or presence (**e**) of a PDE1 inhibitor 8-MM-IBMX. Scale bars indicate 1 s and 200 pA. (**f**) Summary of electrophysiological data showing relation between the strength of conditioning stimulation (duration of depolarization pulses) and the amplitude of potentiation at 30 min. FK506 is an inhibitor of calcineurin. Modified from Kitagawa et al. 2009

Critical Role of GABARAP in the RP Expression and Maintenance

A critical role of a $GABA_AR$-binding protein in the establishment of RP downstream of the CaMKII activation has been clarified (Kawaguchi and Hirano 2007). A key molecule is $GABA_AR$-associated protein (GABARAP), which has binding sites for $GABA_AR$ γ subunit and for tubulin, a constituent of cytoskeletal microtubule (Wang et al. 1999) (Fig. 3.1). When association of GABARAP and $GABA_AR$ γ2 subunit is inhibited by a peptide corresponding to an intracellular region of γ2 subunit, the RP induction is impaired. Importantly, application of the peptide even after the establishment of RP attenuates the once potentiated $GABA_AR$ response, suggesting that interaction of GABARAP and γ2 subunit is required not only for the induction of RP but also for maintenance. In addition, the association of GABARAP with tubulin is required for RP. Either overexpression of GABARAP lacking the region binding to tubulin or pharmacological destruction of microtubule impairs the RP induction. Thus, the interaction of $GABA_AR$-GABARAP-microtubule is critical for RP.

Fluorescence resonance energy transfer (FRET) imaging showed that GABARAP undergoes the sustained structural change in response to depolarization of a Purkinje neuron. This long-lasting conformational change of GABARAP depends on the activity of CaMKII. Further, the single amino acid replacement of GABARAP V33E, which impairs the structural change of GABARAP, suppresses the RP induction (Kawaguchi and Hirano 2007). These results suggest that CaMKII-mediated sustained structural change of GABARAP is essential for RP.

GABARAP plays a role in intracellular trafficking and targeting of $GABA_AR$ toward the plasma membrane through association with γ2 subunit (Kneussel et al. 2000; Kittler et al. 2001; Moss and Smart 2001; Kneussel 2002; Nymann-Andersen et al. 2002; Leil et al. 2004; Luscher and Keller 2004; Chen and Olsen 2007; Kanematsu et al. 2007). Therefore, CaMKII-mediated structural change of GABARAP might play a role in RP through facilitating transportation of $GABA_AR$ to the cell surface. In a hippocampal neuron, inhibitory synaptic potentiation is caused by activation of NMDA-type of glutamate receptors through GABARAP-dependent exocytosis of $GABA_AR$ (Marsden et al. 2007). Another possible role of GABARAP in RP is to enhance the $GABA_AR$ function. GABARAP alters the properties of $GABA_AR$ such as single channel conductance through associating with γ2 subunit (Everitt et al. 2004; Luu et al. 2006). Thus, the depolarization-induced conformational change of GABARAP might facilitate the function of individual $GABA_AR$ through alteration of single channel conductance or open time. Further study is necessary to determine which mechanism works to establish RP.

Conclusion

In this chapter we have described the regulation mechanism of RP, a type of postsynaptic plasticity at GABAergic synapses on a cerebellar Purkinje neuron. Induction of RP is exquisitely regulated by the neuronal activity through complicated interactions of protein kinases and phosphatases, and their activity balance is controlled by Ca^{2+}-activated PDE1, $GABA_BR$ etc. Long lasting activation of CaMKII and sustained alteration of $GABA_AR$-GABARAP-microtubule interaction might cooperatively establish and maintain the potentiation of $GABA_AR$ responsiveness in RP.

References

Aizenman CD, Manis PB, Linden DJ (1998) Polarity of long-term synaptic gain change is related to postsynaptic spike firing at a cerebellar inhibitory synapse. Neuron 21:827–835

Bailey CH, Giustetto M, Huang YY et al. (2000) Is heterosynaptic modulation essential for stabilizing Hebbian plasticity and memory? Nat Rev Neurosci 1:11–20

Bhalla US, Iyengar R (1999) Emergent properties of networks of biological signaling pathways. Science 283:381–387

Brandman O, Meyer T (2008) Feedback loops shape cellular signals in space and time. Science 322:390–395

Brandon N, Jovanovic J, Moss S (2002) Multiple roles of protein kinases in the modulation of γ-aminobutyric $acid_A$ receptor function and cell surface expression. Pharmacol Ther 94:113–122

Chen Z, Olsen RW (2007) $GABA_A$ receptor associated proteins: a key factor regulating $GABA_A$ receptor function. J Neurochem 100:279–294

Duguid IC, Smart TG (2004) Retrograde activation of presynaptic NMDA receptors enhances GABA release at cerebellar interneuron-Purkinje cell synapses. Nat Neurosci 7:525–533

Everitt AB, Luu T, Cromer B et al. (2004) Conductance of recombinant $GABA_A$ channels is increased in cells co-expressing $GABA_A$ receptor-associated protein. J Biol Chem 279:21701–21706

Gaiarsa JL, Caillard O, Ben-Ari Y (2002) Long-term plasticity at GABAergic and glycinergic synapses: mechanisms and functional significance. Trends Neurosci 25:564–570

Giese KP, Fedorov NB, Filipkowski RK et al. (1998) Autophosphorylation at Thr286 of the alpha calcium-calmodulin kinase II in LTP and learning. Science 279:870–873

Hansel C, Linden DJ, D'Angelo E (2001) Beyond parallel fiber LTD: the diversity of synaptic and non-synaptic plasticity in the cerebellum. Nat Neurosci 4:467–475

Hashimoto Y, Sharma RK, Soderling TR (1989) Regulation of Ca^{2+}/calmodulin-dependent cyclic nucleotide phosphodiesterase by the autophosphorylated form of Ca^{2+}/calmodulin-dependent protein kinase II. J Biol Chem 264:10884–10887

Hemmings HC Jr, Greengard P, Tung HY et al. (1984) DARPP-32, a dopamine-regulated neuronal phosphoprotein, is a potent inhibitor of protein phosphatase-1. Nature 310:503–505

Houston CM, Hosie AM, Smart TG (2008) Distinct regulation of β2 and β3 subunit-containing cerebellar synaptic $GABA_A$ receptors by calcium/calmodulin-dependent protein kinase II. J Neurosci 28:7574–7584

Houston CM, Smart TG (2006) CaMK-II modulation of $GABA_A$ receptors expressed in HEK293, NG108-15 and rat cerebellar granule neurons. Eur J Neurosci 24:2504–2514

Ito M (2001) Cerebellar long-term depression: characterization, signal transduction, and functional roles. Physiol Rev 81:1143–1195

Kandel ER (2001) The molecular biology of memory storage: a dialogue between genes and synapses. Science 294:1030–1038

Kanematsu T, Mizokami A, Watanabe K et al. (2007) Regulation of $GABA_A$-receptor surface expression with special reference to the involvement of GABARAP ($GABA_A$ receptor-associated protein) and PRIP (phospholipase C-related, but catalytically inactive protein). J Pharmacol Sci 104:285–292

Kano M (1995) Plasticity of inhibitory synapses in the brain: a possible memory mechanism that has been overlooked. Neurosci Res 21:177–182.
Kano M, Rexhausen U, Dreessen J et al. (1992) Synaptic excitation produces a long-lasting rebound potentiation of inhibitory synaptic signals in cerebellar Purkinje cells. Nature 356:601–604
Kano M, Kano M, Fukunaga K et al. (1996) Ca^{2+}-induced rebound potentiation of γ-aminobutyric acid-mediated currents requires activation of Ca^{2+}/calmodulin-dependent kinase II. Proc Natl Acad Sci USA 93:13351–13356
Kawaguchi S, Hirano T (2000) Suppression of inhibitory synaptic potentiation by presynaptic activity through postsynaptic $GABA_B$ receptors in a Purkinje neuron. Neuron 27:339–347
Kawaguchi S, Hirano T (2002) Signaling cascade regulating long-term potentiation of $GABA_A$ receptor responsiveness in cerebellar Purkinje neurons. J Neurosci 22:3969–3976
Kawaguchi S, Hirano T (2006) Integrin α3β1 suppresses long-term potentiation at inhibitory synapses on the cerebellar Purkinje neuron. Mol Cell Neurosci 31:416–426
Kawaguchi S, Hirano T (2007) Sustained structural change of $GABA_A$ receptor-associated protein underlies long-term potentiation at inhibitory synapses on a cerebellar Purkinje neuron. J Neurosci 27:6788–6799
Kitagawa Y, Hirano T, Kawaguchi S (2009) Prediction and validation of a mechanism to control the threshold for inhibitory synaptic plasticity. Mol Syst Biol 5:280
Kittler JT, Rostaing P, Schiavo G et al. (2001) The subcellular distribution of GABARAP and its ability to interact with NSF suggest a role for this protein in the intracellular transport of $GABA_A$ receptors. Mol Cell Neurosci 18:13–25
Kneussel M (2002) Dynamic regulation of $GABA_A$ receptors at synaptic sites. Brain Res Brain Res Rev 39:74–83
Kneussel M, Haverkamp S, Fuhrmann JC et al. (2000) The γ-aminobutyric acid type A receptor ($GABA_A$R)- associated protein GABARAP interacts with gephyrin but is not involved in receptor anchoring at the synapse. Proc Natl Acad Sci USA 97:8594–8599
Komatsu Y (1994) Age-dependent long-term potentiation of inhibitory synaptic transmission in rat visual cortex. J Neurosci 14:6488–6499
Komatsu Y (1996) $GABA_B$ receptors, monoamine receptors, and postsynaptic inositol trisphosphate-induced Ca^{2+} release are involved in the induction of long-term potentiation at visual cortical inhibitory synapses. J Neurosci 16:6342–6352
Leil TA, Chen ZW, Chang CS et al. (2004) $GABA_A$ receptor- associated protein traffics $GABA_A$ receptors to the plasma membrane in neurons. J Neurosci 24:11429–11438
Lisman JE, Zhabotinsky AM (2001) A model of synaptic memory: a CaMKII/PP1 switch that potentiates transmission by organizing an AMPA receptor anchoring assembly. Neuron 31:191–201
Llano I, Leresche N, Marty A (1991) Calcium entry increases the sensitivity of cerebellar Purkinje cells to applied GABA and decreases inhibitory synaptic currents. Neuron 6:565–574
Luscher B, Keller CA (2004) Regulation of $GABA_A$ receptor trafficking, channel activity, and functional plasticity of inhibitory synapses. Pharm Therap 102:195–221
Luu T, Gage PW, Tierney ML (2006) GABA increases both the conductance and mean open time of recombinant $GABA_A$ channels co-expressed with GABARAP. J Biol Chem 281:35699–35708
Malenka RC, Nocoll RA (1999) Long-term potentiation-a decade of progress? Science 285:1870–1874
Marsden KC, Beattie JB, Friedenthal J et al. (2007) NMDA receptor activation potentiates inhibitory transmission through GABA receptor-associated protein-dependent exocytosis of $GABA_A$ receptors. J Neurosci 27:14326–14337
Marty A, Llano I (1995) Modulation of inhibitory synapses in the mammalian brain. Curr Opin Neurobiol 5:335–341
Miller SG, Kennedy MB (1986) Regulation of brain type II Ca^{2+}/calmodulin- dependent protein kinase by autophosphorylation: a Ca^{2+}-triggered molecular switch. Cell 44:861–870
Mittmann W, Hausser M (2007) Linking synaptic plasticity and spike output at excitatory and inhibitory synapses onto cerebellar Purkinje cells. J Neurosci 27:5559–5570

Moss SJ, Smart TG (1996) Modulation of amino acid-gated ion channels by protein phosphorylation. Int Rev Neurobiol 39:1–52
Moss SJ, Smart TG (2001) Constructing inhibitory synapses. Nat Rev Neurosci 2:240–250
Nusser Z, Hajos N, Somogyi P et al. (1998) Increased number of synaptic $GABA_A$ receptors underlies potentiation at hippocampal inhibitory synapses. Nature 395:172–177
Nymann-Andersen J, Wang H, Chen L et al. (2002) Subunit specificity and interaction domain between $GABA_A$ receptor-associated protein (GABARAP) and $GABA_A$ receptors. J Neurochem 80:815–823
Sugiyama Y, Kawaguchi S, Hirano T (2008) mGluR1-mediated facilitation of long-term potentiation at inhibitory synapses on a cerebellar Purkinje neuron. Eur J Neurosci 27:884–896
Wang H, Bedford FK, Brandon NJ et al. (1999) $GABA_A$-receptor-associated protein links $GABA_A$ receptors and the cytoskeleton. Nature 397:69–72

Chapter 4
Inhibitory Plasticity and Auditory Function

Robert C. Liu

Abstract The mammalian auditory system serves many functions for an organism throughout its life, including the spatial localization of sound sources and the recognition of behaviorally-relevant sounds. The neural circuitry underlying these various functions are not all fully elucidated, but likely involve key contributions from both subcortical as well as cortical auditory areas. While the focus has often been on the role of neural excitation within these areas, there is increasing recognition that neural inhibition and its plasticity can be just as important in shaping the function of auditory circuits.

The purpose of this chapter is to use a few of the known examples of how inhibition and inhibitory plasticity subserve the functional processing of sounds to illustrate both the advances that have been made in terms of understanding mechanisms as well as the open questions remaining. The first example deals with sound localization and the brainstem circuitry that decodes binaural spatial cues. This provides one of the most detailed auditory examples of the function of inhibitory synaptic input, how inhibitory circuitry is modified by activity during development, and what potential cellular mechanisms are critical for this plasticity. The other two examples focus on the neural coding and plasticity of behaviorally relevant sounds at the cortical level. One centers on how inhibition shapes the selectivity of individual neurons for frequency modulated sounds during development, and the other on its hypothesized role in the population representation of a communication call in adults. While the cellular mechanisms underlying inhibitory plasticity are less clear in these latter cases, taken all together these examples demonstrate the importance and pervasiveness of inhibition in the functional processing of sounds.

R.C. Liu (✉)
Department of Biology, Emory University, 1510 Clifton Rd NE, Atlanta, GA, USA
e-mail: robert.liu@emory.edu

M.A. Woodin and A. Maffei (eds.), *Inhibitory Synaptic Plasticity*,
DOI 10.1007/978-1-4419-6978-1_4, © Springer Science+Business Media, LLC 2011

Inhibitory Plasticity Refines Brainstem Sound Localization Circuits

The localization of sounds depends on a variety of acoustic cues from a sound source. Monaural spectral cues arise based on how the outer ear differentially filters sound frequencies from sources at different spatial locations (Rice et al. 1992; Fuzessery 1996). Binaural cues arise due to differences in the same sound detected by the two ears. Because the head casts an acoustic shadow for sounds arising from one side to the other, interaural level differences (ILD) in the sound amplitude are created at the two ears. The delay for sounds to travel across the head also produces an interaural timing difference (ITD) cue. For lateralized sources, the relative importance of these two cues for a given animal depends on the wavelength of the sound compared to its head size. A large head presents a greater obstacle to the propagation of high frequency (small wavelength) rather than low frequency sounds, making ILD particularly effective for the former. For example, in humans, frequencies above a few kilohertz (kHz) can generate an ILD as high as 15 dB, whereas low frequencies <2 kHz (>17 cm wavelength) may only produce a maximum ILD of 5 dB (Wightman and Kistler 1993). At these lower frequencies, ITD is a more effective cue, provided acoustic delays can be faithfully represented in the neural activity arising from the two ears.

The neural activity induced by sounds originates in the inner ear with the mechanotransduction of frequency-specific basilar membrane deflections by inner hair cells arrayed along the organ of Corti (Geisler 1998). Primary auditory nerve fibers (ANF) innervating these hair cells project to the ipsilateral cochlear nucleus; their axons bifurcate to form synapses in its three tonotopically organized divisions, the dorsal, anteroventral (AVCN) and posteroventral cochlear nuclei. For low frequency sounds below a few kHz (Joris et al. 1994), ANFs fire in a phase-locked manner with precise timing (Rose et al. 1967), thereby preserving potential information about acoustic delays between the ears. For higher frequency sounds, phase-locking to each cycle breaks down, but the rate of ANF spiking at the onset and through a sound's duration can nevertheless encode the sound level at that ear (Yates et al. 1985). Many cochelar nucleus neurons preserve these ANF firing patterns (Blackburn and Sachs 1989), relaying the neural activity to higher auditory centers in the brainstem.

This spike timing and rate information from the two ears are first combined at the level of the superior olive. The lateral superior olive (LSO) is believed to mainly decode ILD cues (Tollin and Yin 2002a), while the medial superior olive is sensitive to small ITDs (Yin and Chan 1990). Neurons in both areas are organized along a tonotopic axis (Kandler et al. 2009). For the LSO, frequency-appropriate glutamatergic excitatory inputs mainly arise from spherical bushy cells in the ipsilateral AVCN (Cant and Casseday 1986), while (predominantly) glycinergic inhibitory inputs come from the ipsilateral medial nucleus of the trapezoid body (MNTB), whose neurons are excited by globular bushy cells of the contralateral AVCN (Smith et al. 1991). Hence, LSO neurons receive ipsilateral excitation and contralateral inhibition. On the other hand, MSO neurons receive binaural glutamatergic excitatory and glycinergic inhibitory inputs (Grothe 2003): excitation arrives via spherical bushy cell projections from both AVCNs (Cant and Casseday 1986), while inhibition comes most strongly from the contralaterally-driven MNTB, as well as the ipsilaterally-driven lateral nucleus of the trapezoid body (LNTB), both located ipsilateral to the MSO. For both sound localization nuclei then, the trapezoid body provides an important source of feed-forward inhibition.

These inhibitory inputs play a critical role in forming the sensitivity of LSO and MSO neurons to their respective cues. Focusing first on the LSO, in vivo electrophysiological experiments in adult cats using virtual sound stimuli presented through earphones proved that contralateral inhibition shapes spatial receptive fields (SRF) (Tollin and Yin 2002b). Simulating the binaural acoustic cues arising from sound sources at different azimuths, spike rate responses were generally tuned for ipsilateral locations, and inhibited below the spontaneous level for contralateral locations. However, if the contralateral earphone was silenced so that the LSO only received ipsilateral (excitatory) inputs, firing rates increased and spatial tuning shifted toward the midline. Additional controls made possible by the virtual sound paradigm showed that these SRFs were indeed based primarily on ILD cues for sounds with frequencies near the LSO neuron's characteristic frequency (Tollin and Yin 2002a).

To construct the SRF, the excitatory and inhibitory inputs to an LSO neuron need to be frequency-matched. Indeed, the characteristic frequencies of excitation and inhibition (obtained through a two-tone stimulation paradigm, see later) are nearly perfectly correlated in adult Mongolian gerbils (Sanes and Rubel 1988). Just after hearing onset (which occurs ~postnatal day P12) though, the correlation is much lower and tuning bandwidths of each are much larger (Sanes and Rubel 1988). Evidence now points to a two-step process of functional synaptic silencing and structural pruning in the inhibitory pathway to produce the frequency alignment along the tonotopic axis observed in the adult LSO (Kandler et al. 2009).

Just after birth, individual MNTB axons (in Mongolian gerbils) are already innervating the LSO in a coarse tonotopic fashion, and during the first two weeks of life (before hearing onset), their axonal arborization continues to grow (Sanes and Siverls 1991). Over the same period, the dendritic tree of a LSO neuron is quite broad and may actually be growing, although there is substantial variability from neuron to neuron (Sanes et al. 1992a; Rietzel and Friauf 1998). Surprisingly though, glutamate uncaging across an array of MNTB cells while whole-cell recording from a rat LSO neuron demonstrated a progressive decline over the first postnatal week in the size of the tonotopic MNTB region providing synaptic input (Kim and Kandler 2003). Moreover, the ~75% reduction in the width of functional inputs was accompanied by a 12-fold increase in the conductance evoked by the remaining MNTB inputs, suggesting enhanced specificity for tonotopically-appropriate projections. Hence, despite little evidence for structural pruning prior to hearing onset, functional silencing of synapses from tonotopically-inappropriate MNTB neurons occurs very early in development.

Just after hearing onset, structural plasticity of the inhibitory input becomes evident. MNTB axonal arborizations become more refined as the spread in the width of the bouton distribution and the number of boutons both decrease by ~35% between P12-13 and P18-25 (Sanes and Siverls 1991). Concomitantly, LSO dendritic fields along the tonotopic dimension decrease, as do the number of branches and endpoints (Sanes et al. 1992a; Rietzel and Friauf 1998). Auditory experience (or at least spontaneous activity from auditory afferents) is likely needed for this structural refinement since contralateral cochlear removal at P7, which leads to degeneration of the inputs to MNTB neurons and decreased drive to LSO neurons, impedes the pruning process and can even cause hypertrophic LSO dendritic branching (Sanes et al. 1992b; Sanes and Takacs 1993).

What cellular mechanisms are hypothesized to underlie the refinement of this inhibitory circuitry? Several developmental changes in the nature of the MNTB-LSO inhibition likely play a role. First, immature MNTB cells actually release GABA in addition to glycine, and neurotransmission gradually shifts from the former to the latter over the first 2 weeks (Kotak et al. 1998). During the first postnatal week, both are depolarizing and can elicit action potentials in the LSO (Kandler and Friauf 1995; Kotak et al. 1998; Kullmann and Kandler 2001). Both can increase calcium levels postsynaptically at this age, but they decrease calcium levels after hearing onset (Kullmann et al. 2002). During the early window, calcium responses can be spatially restricted if generated by subthreshold synaptic activity (Kullmann and Kandler 2008). This could drive synapse-specific strength modifications through a variety of calcium-dependent mechanisms (Luscher and Keller 2004), including retrograde $GABA_B$ signaling (Magnusson et al. 2008) or GABA-mediated long-term synaptic depression (Kotak and Sanes 2000; Kotak et al. 2001; Chang et al. 2003). Second, there is evidence that during the first postnatal week, glutamate is also released at the MNTB synaptic terminals and can act on NMDA receptors (Gillespie et al. 2005). These receptors are no longer blocked by magnesium due to the depolarizing action of GABA, and they can induce long term synaptic plasticity even at inhibitory synapses (Gaiarsa et al. 2002). While these mechanisms all act before hearing onset, driven presumably by spontaneous activity (Walsh and McGee 1988; Chang et al. 2003), why structural plasticity is delayed until after hearing onset and which mechanisms are essential for it are not yet fully understood (Kandler et al. 2009).

The importance of inhibition and inhibitory plasticity is also becoming apparent for MSO circuitry, although less is known here than in the LSO case. Inhibitory projections to MSO were identified long ago (Cant and Hyson 1992; Kuwabara and Zook 1992), but their function was not initially understood. The prevailing model at the time for decoding ITD was based on Jeffress's theory of an array of coincidence detector neurons responding maximally to different ITDs by combining excitation arriving via delay lines from the two ears (Jeffress 1948). Evidence for this was found in the nucleus laminaris of birds (Carr and Konishi 1988; Overholt et al. 1992), and the MSO of awake rabbits (Batra et al. 1997), but the model did not always fit for mammals (Grothe 2003; McAlpine and Grothe 2003).

In Mongolian gerbils, which specialize in low-frequency hearing, ITDs eliciting the highest spike rates (best ITD) were outside of the physiologically relevant range (±120 μs) expected for the its head size (Brand et al. 2002). Instead, the slopes of the ITD functions generally fell across this range, leading to the hypothesis that ITD in low-frequency hearing mammals is encoded by the relative firing of a population of MSO neurons rather than a labeled line from individual tuned detectors (McAlpine and Grothe 2003). A now classic in vivo experiment in adult Mongolian gerbils revealed that inhibitory inputs are essential in shaping this ITD tuning (Brand et al. 2002). Iontophoretically applying the glycine antagonist strychnine during extracellular electrophysiology demonstrated that inhibition positions the slope of the ITD tuning curve within the physiologically relevant range and the peak on the contralaterally leading side. Without it, best ITDs are shifted toward zero, with the same firing rate ambiguously corresponding to different ITDs within the physiologically relevant range. Modeling suggested that these results could be explained by a strong contralateral inhibition arriving just before contralateral excitation, which would effectively suppress ipsilateral excitation from sounds originating ipsilaterally and produce ITD tuning peaked for contralaterally leading sounds (Brand et al. 2002). This fits with the observed timing of excitation and inhibition in the bat MSO (Grothe et al. 1997), and the greater strength of contralateral compared to ipsilateral inhibition (Grothe and Sanes 1993; Grothe and Park 1998).

What mechanisms enable this strong, rapid inhibition? First, MNTB neurons receive synaptic input from AVCN globular bushy neurons via a single calyx of Held synaptic terminal that wraps around the cell body (Satzler et al. 2002). This is thought to be one of the most secure synapses in the mammalian nervous system, and it activates the MNTB quickly and with high temporal precision (Smith et al. 1998). Second, in adult animals that specialize in low-frequency hearing (e.g. Mongolian gerbil), the MNTB inhibitory input onto an MSO neuron is heavily concentrated on the soma, probably allowing for faster kinetics of inhibition (Kapfer et al. 2002). Importantly, this localized distribution appears to be a result of experience-dependent refinement. The distribution of glycinergic inputs is initially diffuse across an MSO neuron's soma and dendrites before hearing onset (P10). Removing the cochlea at P7 and thus "monauralizing" the input leaves the distribution in this juvenile state (Kapfer et al. 2002). Moreover, raising intact animals in omnidirectional white noise, which is thought to diminish sound localization cues (Withington-Wray et al. 1990), still leaves significantly more glycinergic inputs on distal dendrites than in normal adults (Kapfer et al. 2002), and produces ITD tuning curves (measured downstream of MSO) more consistent with juveniles (Seidl and Grothe 2005). Finally, this latter manipulation also impedes the normal, post-hearing-onset pruning of the MNTB axonal projections onto MSO neurons (Werthat et al. 2008). These studies all indicate that inhibitory MNTB projection neurons, which send axon collaterals to both LSO and MSO (Banks and Smith 1992), undergo substantial plasticity in their functional as well as anatomical connections during development. However, whether the same cellular mechanisms for refinement of LSO synapses also apply to the MSO remain to be investigated.

Development of Inhibitory Sidebands Tunes Cortical Selectivity for FM Sweeps

Besides the spatial localization of sound sources, a second critical function of the auditory system deals with the processing of behaviorally relevant sounds such as communication calls. Individual neurons in the mammalian auditory system can be selective to acoustic features within natural calls, but populations of neurons are believed necessary to represent each call fully (Wang 2000). One such feature occurring in the calls of many species, including humans, is the frequency-modulated (FM) sweep (Lindblom and Studdert-Kennedy 1967; Pola and Snowdon 1975; Kanwal et al. 1994; Liu et al. 2006; Razak and Fuzessery 2006). Its behavioral relevance is demonstrated by the fact that it can be used to discriminate different phonemes in human speech (Lindblom and Studdert-Kennedy 1967). In constructing a neuron's selectivity for FM features like sweep direction and rate, it is now known that at the level of the auditory cortex, inhibition plays a major role (Zhang et al. 2003; Razak and Fuzessery 2006) and is subject to experience-dependent plasticity during development (Razak and Fuzessery 2007; Razak et al. 2008).

The function and development of inhibition for FM selectivity was recently detailed in a series of studies on the anesthetized pallid bat. This species uses downward ultrasonic FM sweeps (e.g. 60–30 kHz) for echolocation, and the majority of high frequency cortical neurons are selectively excited by these calls and tuned around behaviorally-relevant values of the sweep direction and rate (Razak and Fuzessery 2002, 2006). The excitatory frequency tuning curves of such cortical neurons are usually flanked by inhibitory sidebands (Suga 1965). These sidebands are detected by a two-tone paradigm wherein a short "probe" tone guaranteed to excite the neuron (usually presented ~10–20 dB above the neuron's threshold at its characteristic frequency) is paired with a second tone that varies in frequency and relative timing. If the probe response is suppressed to some criterion level, then the frequency and timing of the second tone fall within the inhibitory sideband. In adult bats, these sidebands are asymmetric, with low frequency inhibition (LFI) having a broader bandwidth and earlier arrival time than high frequency inhibition (HFI) (Razak and Fuzessery 2006). This structure can explain both the FM rate and direction selectivity of cortical neurons. Intuitively, in order to selectively respond to a particular rate of downward FM (e.g. ranging from 60 to 30 kHz within 10 ms for a 3 kHz/ms sweep rate), the inhibitory input triggered when the initial high frequencies in that sweep enter the HFI sideband should not suppress the corresponding excitatory input. Hence, the frequency width of the high frequency edge of the HFI compared to the high frequency edge of the excitatory tuning curve, divided by the arrival time of the inhibitory input relative to the excitatory input, can be used to predict a neuron's cutoff rate for responding, and generally agrees quite well with observed cutoffs (Razak and Fuzessery 2006). Similarly, to achieve selectivity for downward rather than upward FM, the LFI should arrive before excitation for upward sweeps for a wide range of rates. This requires both a broadly tuned and early inhibitory input, which has been found to correctly predict the direction selectivity index (Zhang et al. 2003) of cortical neurons (Razak and Fuzessery 2006).

Developmentally, adult FM rate selectivity and the corresponding structure of HFI are already in place in 2 week old bats, at the time when they are just beginning to hear the frequencies present in echolocation calls (Razak and Fuzessery 2007). However, FM direction selectivity and LFI properties change slowly over time, and begin approaching adult parameters only by around 12 weeks of age (Razak and Fuzessery 2007), a time point just beyond when the pups are weaned and are using the calls to navigate (Brown 1976). Both types of FM selectivity require some experience in order to reach the adult condition. This was discovered by paralyzing the laryngeal muscles of pups younger than 2 weeks old to alter the subsequent production of echolocation vocalizations; calls were significantly lower in the highest achievable frequency and slower in sweep rate (Razak et al. 2008). While FM rate selectivity was present at P30 and P90 in these experimental (EXP) animals, as in intact isolation-reared (CTRL) and normal (NORM) animals, the percentage of neurons showing rate selectivity was significantly lower in the former, suggesting that experience helps *maintain* rate selectivity in the cortex (Razak et al. 2008). This corresponded to a larger percentage of neurons that did not show HFI. On the other hand, experience appears to be necessary for both the development and maintenance of direction selectivity since EXP animals started out with similarly poor direction selectivity as CTRL and NORM animals at P30, but actually got worse by P90 (Razak et al. 2008). This corresponded to significantly less LFI, as well as later LFI arrival times in P90 EXP animals. Taken together, these studies demonstrate a role for experience-dependent plasticity in inhibitory circuitry to shape the selectivity for vocalization features in individual cortical neurons. The cellular mechanisms by which this occurs, however, remain to be discovered.

Inhibitory Plasticity in the Population Representation of a Communication Call

In addition to affecting the selectivity of individual neurons to features of natural vocalizations, inhibitory mechanisms may also be relevant for representing communication sounds at the neural population level. This possibility was recently revealed in a study looking at how the auditory cortical representation of a specific call changed as this call became behaviorally relevant to mice (Galindo-Leon et al. 2009). The vocalization is the ultrasonic isolation call (~65 kHz) of mouse pups (Liu et al. 2003), which elicits a directed search and retrieval behavior in mothers (Sewell 1970; Smith 1976; Haack et al. 1983). On the other hand, virgin females without pup exposure do not recognize the calls as behaviorally relevant, and do not preferentially approach them over a neutral ultrasound (Ehret et al. 1987). This has opened the possibility to explore how the adult auditory system may undergo plasticity in natural communication contexts (Miranda and Liu 2009). Indeed, a series of studies in the anesthetized mouse demonstrated that multiunit auditory cortical activity of mothers compared to virgins is better adapted to detect and discriminate individual pup calls (Liu and Schreiner 2007) as well as to follow trains of ultrasonic pup calls (Liu et al. 2006).

The development of an awake, head-restrained auditory cortex electrophysiology preparation in the mouse has now extended those studies and demonstrated some surprising results. First, whereas the anesthetized multiunit studies emphasized excitatory neural responses, (Galindo-Leon et al. 2009) found that around half the single units (SU) which responded to calls did so in a purely inhibitory manner, in both mothers and virgins. Second, features of this inhibition were found to be markedly changed in mothers: the duration of call-inhibited SUs was longer, its onset slightly earlier, and the consistency and depth of its spiking suppression greater (Fig. 4.1a). These are consistent with the idea that inhibition itself is stronger in

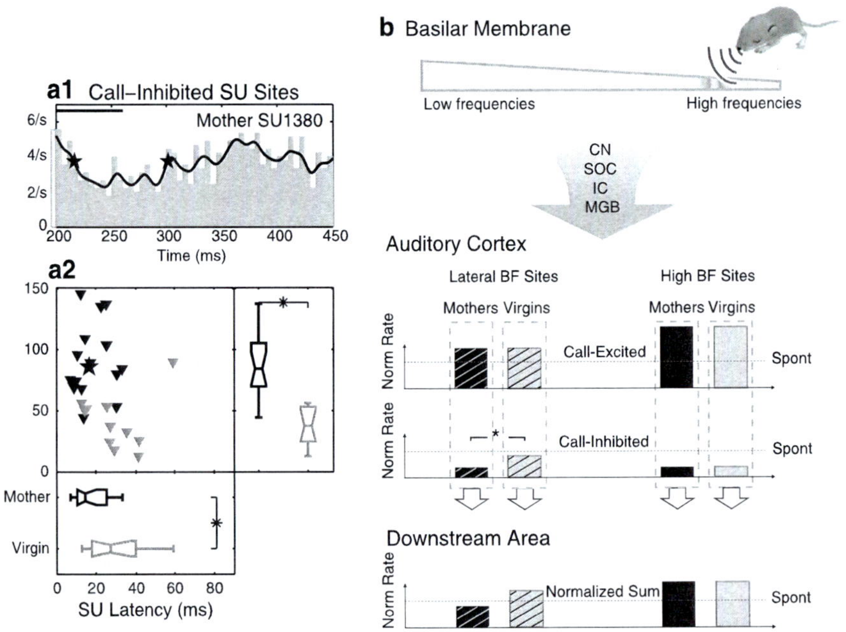

Fig. 4.1 Inhibitory plasticity in the response to an ultrasonic pup call between virgins and mothers may improve the neural contrast in the population representation of the calls. (**a1**) Example post stimulus time histogram (*gray*) of responses to a collection of pup calls (presented during *black* horizontal bar). The timing and duration of inhibition are marked by stars. (**a2**) Group comparison of call-evoked inhibitory onset and durations. (**b**). Hypothesized model to enhance a pup call's neural contrast. Activity from call-excited SUs in both mothers and virgins are comparable, but significantly lower in mothers for call-inhibited SUs in the lateral band. A downstream area that integrates the contributions from both call-excited and call-inhibited SUs in a frequency-band specific manner has a greater contrast between the two frequency regions. CN: cochlear nucleus; SOC: superior olivary complex nuclei; IC: inferior colliculus; MGB: medial geniculate body. Based on Neuron, vol. 62, E. E. Galindo-Leon, F. G. Lin, R. C. Liu, "Inhibitory Plasticity in a Lateral Band Improves Cortical Detection of Natural Vocalizations," 705–716, (2009), with permission from Elsevier

mothers. Interestingly, this difference appeared not only in the spiking activity, but also in the collective subthreshold activity of neurons. Local field potentials (LFP) recorded around inhibited SUs showed significant differences between virgins and mothers as well, primarily in the trial-by-trial variability of the LFP shape. Although the LFP is often thought to be a coarse neural population measure (Logothetis 2003), it can be strongly correlated with the membrane activity of individual neurons (Poulet and Petersen 2008), presumably because of synchronized membrane activity across a neural population (DeWeese and Zador 2006). The fact that the two animal groups also differed in LFP variability around inhibited SUs then suggests that a network-level plasticity involving inhibition.

Finally, this inhibitory plasticity was not found to be equally robust across the auditory cortex. The strongest changes between virgins and mothers were actually at neural sites tuned to frequencies *below* that of the pup calls. Assuming that the SUs recorded by fairly high impedance tungsten electrodes are primarily pyramidal cells, this implies that areas of the auditory cortex that should not respond to pup calls because they represent lower frequencies are being better inhibited by these calls (presented at moderately loud sound levels) in mothers compared to virgins. Moreover, this improved call-evoked inhibition was relatively specific among these so-called lateral band sites for the high ultrasonic frequencies found in pup calls. This suggests that in mothers, a population level adaptation which occurs when these isolation calls become behaviorally relevant is an improved contrast in neural activity between the high frequency and lateral band frequency regions of the auditory cortex (Fig. 4.1b). This may permit a more robust neural population representation of these calls in the presence of low frequency background noise. This hypothesis remains to be tested, as do the possible mechanisms by which these changes occurred. In particular, whether attentional, hormonal or experience-dependent mechanisms drive this inhibitory plasticity has yet to be explored.

Summary

The examples of inhibitory plasticity described here span a range of auditory functions and processing levels. Whether for sound localization or identification, in the brainstem or the cortex, inhibition helps to shape how auditory neurons respond to behaviorally relevant cues in sounds. Functionally relevant plasticity in this inhibition occurs not only during developmental refinement, as in the first two examples, but also in adulthood during large physiological changes such as motherhood. In some cases, such as the sound localization circuitry in the superior olive, our knowledge has advanced sufficiently to describe not only the phenomenology, but also begin to tease apart potential mechanisms, such as GABA depolarization and NMDA receptor-mediated plasticity. Indeed, the effects of GABA antagonists in the cortex during development of FM selectivity would make a nice mechanistic extension of that work. In the last case, more phenomenology may still be necessary to fully characterize changes in inhibition, but mechanistic studies addressing the role of hormones or experience will certainly help as well.

References

Banks MI, Smith PH (1992) Intracellular recordings from neurobiotin-labeled cells in brain slices of the rat medial nucleus of the trapezoid body. J Neurosci 12:2819–2837.

Batra R, Kuwada S, Fitzpatrick DC (1997) Sensitivity to interaural temporal disparities of low- and high-frequency neurons in the superior olivary complex. II. Coincidence detection. J Neurophysiol 78:1237–1247.

Blackburn CC, Sachs MB (1989) Classification of unit types in the anteroventral cochlear nucleus: PST histograms and regularity analysis. J Neurophysiol 62:1303–1329.

Brand A, Behrend O, Marquardt T, McAlpine D, Grothe B (2002) Precise inhibition is essential for microsecond interaural time difference coding. Nature 417:543–547.

Brown P (1976) Vocal communication in the pallid bat, Antrozous pallidus. Z Tierpsychol 41:34–54.

Cant NB, Casseday JH (1986) Projections from the anteroventral cochlear nucleus to the lateral and medial superior olivary nuclei. J Comp Neurol 247:457–476.

Cant NB, Hyson RL (1992) Projections from the lateral nucleus of the trapezoid body to the medial superior olivary nucleus in the gerbil. Hear Res 58:26–34.

Carr CE, Konishi M (1988) Axonal delay lines for time measurement in the owl's brainstem. Proc Natl Acad Sci U S A 85:8311–8315.

Chang EH, Kotak VC, Sanes DH (2003) Long-term depression of synaptic inhibition is expressed postsynaptically in the developing auditory system. J Neurophysiol 90:1479–1488.

DeWeese MR, Zador AM (2006) Non-Gaussian membrane potential dynamics imply sparse, synchronous activity in auditory cortex. J Neurosci 26:12206–12218.

Ehret G, Koch M, Haack B, Markl H (1987) Sex and parental experience determine the onset of an instinctive behavior in mice. Naturwissenschaften 74:47.

Fuzessery ZM (1996) Monaural and binaural spectral cues created by the external ears of the pallid bat. Hear Res 95:1–17.

Gaiarsa JL, Caillard O, Ben-Ari Y (2002) Long-term plasticity at GABAergic and glycinergic synapses: mechanisms and functional significance. Trends Neurosci 25:564–570.

Galindo-Leon EE, Lin FG, Liu RC (2009) Inhibitory plasticity in a lateral band improves cortical detection of natural vocalizations. Neuron 62:705–716.

Geisler CD (1998) From sound to synapse : physiology of the mammalian ear. New York: Oxford University Press.

Gillespie DC, Kim G, Kandler K (2005) Inhibitory synapses in the developing auditory system are glutamatergic. Nat Neurosci 8:332–338.

Grothe B (2003) New roles for synaptic inhibition in sound localization. Nat Rev Neurosci 4:540–550.

Grothe B, Sanes DH (1993) Bilateral inhibition by glycinergic afferents in the medial superior olive. J Neurophysiol 69:1192–1196.

Grothe B, Park TJ (1998) Sensitivity to interaural time differences in the medial superior olive of a small mammal, the Mexican free-tailed bat. J Neurosci 18:6608–6622.

Grothe B, Park TJ, Schuller G (1997) Medial superior olive in the free-tailed bat: response to pure tones and amplitude-modulated tones. J Neurophysiol 77:1553–1565.

Haack B, Markl H, Ehret G (1983) Sound communication between parents and offspring. In: The auditory psychobiology of the mouse (Willott JF, ed), pp 57–97. Springfield, IL: Charles C. Thomas.

Jeffress LA (1948) A place theory of sound localization. J Comp Physiol Psychol 41:35–39.

Joris PX, Carney LH, Smith PH, Yin TC (1994) Enhancement of neural synchronization in the anteroventral cochlear nucleus. I. Responses to tones at the characteristic frequency. J Neurophysiol 71:1022–1036.

Kandler K, Friauf E (1995) Development of glycinergic and glutamatergic synaptic transmission in the auditory brainstem of perinatal rats. J Neurosci 15:6890–6904.

Kandler K, Clause A, Noh J (2009) Tonotopic reorganization of developing auditory brainstem circuits. Nat Neurosci 12:711–717.

Kanwal JS, Matsumura S, Ohlemiller K, Suga N (1994) Analysis of acoustic elements and syntax in communication sounds emitted by mustached bats. J Acoust Soc Am 96:1229–1254.
Kapfer C, Seidl AH, Schweizer H, Grothe B (2002) Experience-dependent refinement of inhibitory inputs to auditory coincidence-detector neurons. Nat Neurosci 5:247–253.
Kim G, Kandler K (2003) Elimination and strengthening of glycinergic/GABAergic connections during tonotopic map formation. Nat Neurosci 6:282–290.
Kotak VC, Sanes DH (2000) Long-lasting inhibitory synaptic depression is age- and calcium-dependent. J Neurosci 20:5820–5826.
Kotak VC, DiMattina C, Sanes DH (2001) GABA(B) and Trk receptor signaling mediates long-lasting inhibitory synaptic depression. J Neurophysiol 86:536–540.
Kotak VC, Korada S, Schwartz IR, Sanes DH (1998) A developmental shift from GABAergic to glycinergic transmission in the central auditory system. J Neurosci 18:4646–4655.
Kullmann PH, Kandler K (2001) Glycinergic/GABAergic synapses in the lateral superior olive are excitatory in neonatal C57Bl/6J mice. Brain Res Dev Brain Res 131:143–147.
Kullmann PH, Kandler K (2008) Dendritic Ca2+ responses in neonatal lateral superior olive neurons elicited by glycinergic/GABAergic synapses and action potentials. Neuroscience 154:338–345.
Kullmann PH, Ene FA, Kandler K (2002) Glycinergic and GABAergic calcium responses in the developing lateral superior olive. Eur J Neurosci 15:1093–1104.
Kuwabara N, Zook JM (1992) Projections to the medial superior olive from the medial and lateral nuclei of the trapezoid body in rodents and bats. J Comp Neurol 324:522–538.
Lindblom BEF, Studdert-Kennedy M (1967) On the role of formant transitions in vowel recognition. J Acoust Soc Am 42:830–843.
Liu RC, Schreiner CE (2007) Auditory cortical detection and discrimination correlates with communicative significance. PLoS Biol 5:e173.
Liu RC, Linden JF, Schreiner CE (2006) Improved cortical entrainment to infant communication calls in mothers compared with virgin mice. Eur J Neurosci 23:3087–3097.
Liu RC, Miller KD, Merzenich MM, Schreiner CE (2003) Acoustic variability and distinguishability among mouse ultrasound vocalizations. J Acoust Soc Am 114:3412–3422.
Logothetis NK (2003) The underpinnings of the BOLD functional magnetic resonance imaging signal. J Neurosci 23:3963–3971.
Luscher B, Keller CA (2004) Regulation of GABAA receptor trafficking, channel activity, and functional plasticity of inhibitory synapses. Pharmacol Ther 102:195–221.
Magnusson AK, Park TJ, Pecka M, Grothe B, Koch U (2008) Retrograde GABA signaling adjusts sound localization by balancing excitation and inhibition in the brainstem. Neuron 59:125–137.
McAlpine D, Grothe B (2003) Sound localization and delay lines–do mammals fit the model? Trends Neurosci 26:347–350.
Miranda JA, Liu RC (2009) Dissecting natural sensory plasticity: hormones and experience in a maternal context. Hear Res 252:21–28.
Overholt EM, Rubel EW, Hyson RL (1992) A circuit for coding interaural time differences in the chick brainstem. J Neurosci 12:1698–1708.
Pola YV, Snowdon CT (1975) The vocalizations of pygmy marmosets (Cebuella pygmaea). Anim Behav 23:826–842.
Poulet JF, Petersen CC (2008) Internal brain state regulates membrane potential synchrony in barrel cortex of behaving mice. Nature 454:881–885.
Razak KA, Fuzessery ZM (2002) Functional organization of the pallid bat auditory cortex: emphasis on binaural organization. J Neurophysiol 87:72–86.
Razak KA, Fuzessery ZM (2006) Neural mechanisms underlying selectivity for the rate and direction of frequency-modulated sweeps in the auditory cortex of the pallid bat. J Neurophysiol 96:1303–1319.
Razak KA, Fuzessery ZM (2007) Development of inhibitory mechanisms underlying selectivity for the rate and direction of frequency-modulated sweeps in the auditory cortex. J Neurosci 27:1769–1781.

Razak KA, Richardson MD, Fuzessery ZM (2008) Experience is required for the maintenance and refinement of FM sweep selectivity in the developing auditory cortex. Proc Natl Acad Sci U S A 105:4465–4470.
Rice JJ, May BJ, Spirou GA, Young ED (1992) Pinna-based spectral cues for sound localization in cat. Hear Res 58:132–152.
Rietzel HJ, Friauf E (1998) Neuron types in the rat lateral superior olive and developmental changes in the complexity of their dendritic arbors. J Comp Neurol 390:20–40.
Rose JE, Brugge JF, Anderson DJ, Hind JE (1967) Phase-locked response to low-frequency tones in single auditory nerve fibers of the squirrel monkey. J Neurophysiol 30:769–793.
Sanes DH, Rubel EW (1988) The ontogeny of inhibition and excitation in the gerbil lateral superior olive. J Neurosci 8:682–700.
Sanes DH, Siverls V (1991) Development and specificity of inhibitory terminal arborizations in the central nervous system. J Neurobiol 22:837–854.
Sanes DH, Takacs C (1993) Activity-dependent refinement of inhibitory connections. Eur J Neurosci 5:570–574.
Sanes DH, Song J, Tyson J (1992a) Refinement of dendritic arbors along the tonotopic axis of the gerbil lateral superior olive. Brain Res Dev Brain Res 67:47–55.
Sanes DH, Markowitz S, Bernstein J, Wardlow J (1992b) The influence of inhibitory afferents on the development of postsynaptic dendritic arbors. J Comp Neurol 321:637–644.
Satzler K, Sohl LF, Bollmann JH, Borst JG, Frotscher M, Sakmann B, Lubke JH (2002) Three-dimensional reconstruction of a calyx of Held and its postsynaptic principal neuron in the medial nucleus of the trapezoid body. J Neurosci 22:10567–10579.
Seidl AH, Grothe B (2005) Development of sound localization mechanisms in the mongolian gerbil is shaped by early acoustic experience. J Neurophysiol 94:1028–1036.
Sewell GD (1970) Ultrasonic communication in rodents. Nature 227:410.
Smith JC (1976) Responses of adult mice to models of infant calls. J Comp Physiol Psychol 90:1105–1115.
Smith PH, Joris PX, Yin TC (1998) Anatomy and physiology of principal cells of the medial nucleus of the trapezoid body (MNTB) of the cat. J Neurophysiol 79:3127–3142.
Smith PH, Joris PX, Carney LH, Yin TCT (1991) Projections of physiologically characterized globular bushy cell axons from the cochlear nucleus of the cat. J Comp Neurol 304:387–407.
Suga N (1965) Functional properties of auditory neurones in the cortex of echo-locating bats. J Physiol 181:671–700.
Tollin DJ, Yin TC (2002a) The coding of spatial location by single units in the lateral superior olive of the cat. II. The determinants of spatial receptive fields in azimuth. J Neurosci 22:1468–1479.
Tollin DJ, Yin TC (2002b) The coding of spatial location by single units in the lateral superior olive of the cat. I. Spatial receptive fields in azimuth. J Neurosci 22:1454–1467.
Walsh EJ, McGee J (1988) Rhythmic discharge properties of caudal cochlear nucleus neurons during postnatal development in cats. Hear Res 36:233–247.
Wang X (2000) On cortical coding of vocal communication sounds in primates. Proc Natl Acad Sci U S A 97:11843–11849.
Werthat F, Alexandrova O, Grothe B, Koch U (2008) Experience-dependent refinement of the inhibitory axons projecting to the medial superior olive. Dev Neurobiol 68:1454–1462.
Wightman FL, Kistler DJ (1993) Sound Localization. In: Human Psychophysics (Yost WA, Popper AN, Fay RR, eds), pp 155–192. New York, NY: Springer.
Withington-Wray DJ, Binns KE, Dhanjal SS, Brickley SG, Keating MJ (1990) The maturation of the superior collicular map of auditory space in the guinea pig is disrupted by developmental auditory deprivation. Eur J Neurosci 2:693–703.
Yates GK, Robertson D, Johnstone BM (1985) Very rapid adaptation in the guinea pig auditory nerve. Hear Res 17:1–12.
Yin TC, Chan JC (1990) Interaural time sensitivity in medial superior olive of cat. J Neurophysiol 64:465–488.
Zhang LI, Tan AY, Schreiner CE, Merzenich MM (2003) Topography and synaptic shaping of direction selectivity in primary auditory cortex. Nature 424:201–205.

Part II
Addiction

Chapter 5
VTA GABAergic Plasticity: An Inhibitory Synaptic Model of Drug Addiction

Fereshteh S. Nugent

Abstract There is now compelling evidence suggesting that addiction is a pathological form of habit-based learning of the brain that involves drug-induced synaptic plasticity in addiction-related areas of the brain including the ventral tegmental area (VTA). Fortunately, over the last decade, tremendous progress has been made in the identification of neuroplastic changes in the relevant neural circuits involved in the development and maintenance of addiction using "synaptic plasticity models". The current model of addiction supports the idea that the VTA is the major starting point of addiction-associated plasticity of the brain in response to drugs of abuse. While synaptic plasticity at excitatory synapses is well-studied and is correlated with addiction, the role of synaptic plasticity at inhibitory synapses is less well understood. However now there is a growing interest in characterizing and uncovering the underlying mechanisms of these forms of inhibitory plasticity and their link to different aspects of brain function, including the development of addictive behaviors. In this chapter, I will provide a brief synopsis of some forms of synaptic plasticity associated with addiction found at inhibitory GABAergic synapses in the VTA.

Introduction

Drug addiction, a chronic and often relapsing disease, costs the U.S. economy more than half a trillion dollars per year. It is characterized by compulsive drug taking and drug seeking behaviors and a high incidence of relapse to drug use (Koob and Le Moal 1997, 2005). Addiction is especially prevalent among children and teens who are prone to experiment with drugs (Volkow and Li 2005). According to the 2007 National Survey on Drug Use and Health (NSDUH), approximately 3.8 million

F.S. Nugent (✉)
Uniformed Services University of the Health Sciences, Department of Pharmacology, Bethesda, MD 20814, USA
e-mail: fereshteh.nugent@usuhs.mil

M.A. Woodin and A. Maffei (eds.), *Inhibitory Synaptic Plasticity*,
DOI 10.1007/978-1-4419-6978-1_5,

Americans age 12 or older have tried heroin, the most powerful and widely abused opiate, at least once. Treatment options for drug addiction are limited, and even if drug use is stopped, protracted withdrawal, chronic craving and relapse are critical problems. Given these daunting statistics, there is a clear need to understand the basis of drug addiction in order to minimize its incidence and improve prospects for its successful treatment.

Fortunately, over the last decade, tremendous progress has been made in the identification of neuroplastic changes in the neural circuits involved in the development and maintenance of addiction using "synaptic plasticity models" (Kauer and Malenka 2007). Since the first experimental demonstration of long-term potentiation (LTP), one of the most heavily studied forms of synaptic plasticity, by Bliss and Lomo 1973, there has been an exciting explosion of literature on synaptic plasticity ascribing its role in the establishment of diverse experience-dependent behaviors under physiological as well as pathological conditions similar to those seen in drug addiction.

There is now a developing consensus that the neuroplasticity of the mesolimbic dopamine system originating from the ventral tegmental area (VTA) in response to drugs of abuse is the first step in the establishment of the pathological habit-based learning of addiction by the brain. The current model of addiction theories is that the process of addiction may begin with synaptic plasticity in the VTA after only one exposure to the drug before transferring to the nucleus accumbens (NAc, also known as ventral striatum), and subsequently engaging the other major memory circuits such as the dorsal striatum (caudate and putamen), orbitofrontal cortex, prefrontal cortex (PFC), amygdala and hippocampus to form the compulsive drug-seeking and drug-taking behaviors of addicts even months after cessation of drug use and drug withdrawal (Koob et al. 2009a).

Most studies on the synaptic plasticity associated with addiction have focused on excitatory synapses, while only a few examples of GABAergic inhibitory plasticity have been described. In contrast to the highly localized action of glutamatergic plasticity, plasticity involving GABAergic interneurons and synapses has the potential to influence both neuronal output and network functions as well as the synaptic plasticity of glutamatergic synapses (i.e., metaplasticity) (McBain et al. 2009; Nugent and Kauer 2008). The role of GABAergic plasticity in different brain processes has received less attention; however now there is a growing interest in characterizing and uncovering the underlying mechanisms of these forms of inhibitory plasticity and their link to different aspects of brain function, including the development of addictive behaviors.

This chapter will provide a brief synopsis of some forms of synaptic plasticity associated with addiction found at inhibitory GABAergic synapses in the VTA. An endocannabinoid (eCB)-mediated long-term depression (LTD) of excitatory synapses in dorsal striatum was first described elegantly by Lovinger's group (Gerdeman et al. 2002) and now is presumed as a common theme throughout the brain expressed by both excitatory and inhibitory synapses and this LTD can result in persistent changes to both neuronal networks and behavior (Chevaleyre et al. 2007). Since the eCB-LTD has been the topic of several recent extensive reviews, I will summarize it more briefly here, focusing on an inhibitory form of this plasticity associated with cocaine abuse in the VTA. For a more comprehensive discussion on the eCBs and eCB-mediated short-term and long-term plasticity, see the following reviews (Lovinger 2008; Heifets and Castillo 2009; Chevaleyre et al. 2006; Gerdeman and Lovinger 2003).

Neural Reward Pathways and Neuroanatomy of the VTA

Recent addiction studies have focused on the dysregulation of brain reward circuitry (Koob and Le Moal 2001; Betz et al. 2000; Wise 2002). The major components of the neural reward circuitry are contained within the ventral tegmental area (VTA), nucleus accumbens (NAc) and prefrontal cortex (PFC). A simplified circuit diagram is shown in Fig. 5.1. However, memory-associated structures such as dorsal striatum, amygdala, and hippocampus are also considered as key elements of the

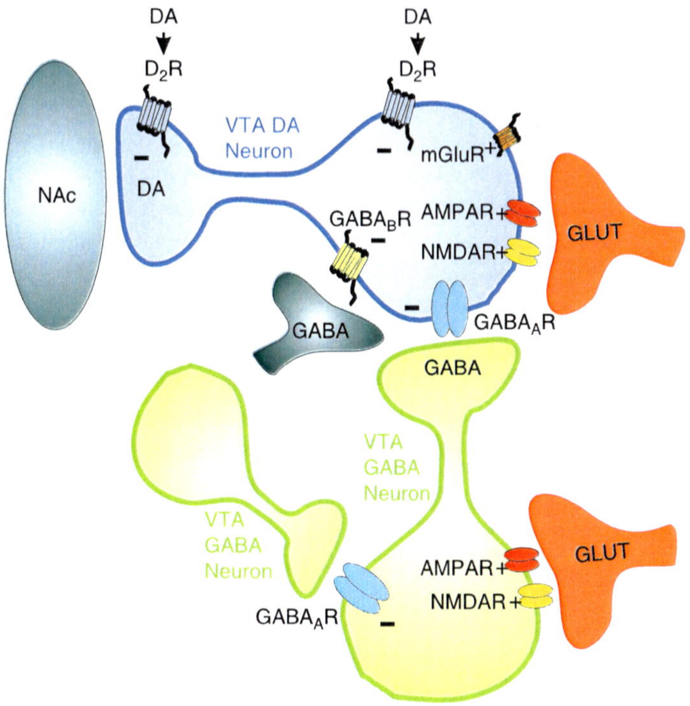

Fig. 5.1 Mesoaccumbens DA circuit diagram focusing on excitatory glutamatergic (+) and inhibitory GABAergic (–) transmissions in the VTA. Locally released DA in the VTA can diffuse many microns from the site of release, binding to the D2 subtype of DA receptor expressed by neighboring DA neurons and inhibits them. Drug-induced plasticity at $GABA_A$ synapses is the focus of this chapter

neurocircuitry of addiction (Koob et al. 2009a). Dopamine (DA) release from DA neurons in the VTA codes for reward and reward prediction (Schultz 1997). DA neuron activity increases during natural pleasurable stimuli such as food, sex, etc., and decreases in response to aversive stimuli such as noxious stimuli (Liu et al. 2008; Ungless et al. 2004).

DA neurons in the VTA receive both excitatory (glutamatergic) and inhibitory (GABAergic) synaptic inputs. The balance of these positive and negative inputs one of the determinants of DA release. The major excitatory inputs to the VTA arise from a direct glutamatergic input from the PFC (Sesack and Pickel 1992; Carr and Sesack 2000; Christie et al. 1985), while subcortical inputs from the laterodorsal tegmental nucleus, lateral hypothalamus, and bed nucleus of the stria terminalis provide a mixture of glutamatergic and what are most likely GABAergic and/or peptidergic projections (Georges et al. 2001; Semba and Fibiger 1992; Garzon et al. 1999). Glutamatergic afferents in the VTA stimulate postsynaptic α-amino-3-hydroxy-5-methyl-4-isoxazolepropionic acid receptors (AMPAR), N-methyl-D-aspartic acid receptors (NMDAR) and metabotropic glutamate receptors (mGluRs) (Bonci and Malenka 1999; Wang and French 1993a, b; Jones and Kauer 1999; Fiorillo and Williams 1998), increasing the firing rates of DA neurons and promoting burst firing *in vivo* (Johnson et al. 1992a; Chergui et al. 1993; Murase et al. 1993). The primary inhibitory afferents to the VTA include the NAc, the ventral pallidum, and the VTA GABAergic interneurons (Bonci and Williams 1996; Johnson et al. 1992b; Fields et al. 2007; Omelchenko and Sesack 2009). Similar to its ability in other brain regions, GABA (γ-aminobutyric acid) in the VTA mediates synaptic inhibition through activation of $GABA_A$ ionotropic receptors (which are chloride channels), and of $GABA_B$ G protein-coupled receptors (which are linked to K^+ channels) (Churchill et al. 1992). VTA neurons are tonically inhibited (hyperpolarized) by endogenous GABA through activation of these receptors, generating a fast inhibitory postsynaptic potential/current (IPSP/IPSC) and a slow IPSP/IPSC, respectively.

Drugs of Abuse and the Neural Reward Pathway

In spite of their different mechanisms of action, all drugs of abuse share three common features (Wolf 2003; Koob 1992). First, they all activate the mesocorticolimbic dopaminergic system - the reward pathway that increases DA levels in the NAc and PFC (Hyman and Malenka 2001; Di Chiara and Imperato 1988; Koob et al. 1989). Second, all drugs of abuse induce negative affective states of withdrawal through inhibition of VTA DA cell activity and consequently low levels of DA in the NAc (Koob and Le Moal 1997, 2005; Koob 1992, 2008; Koob et al. 1989, 2009b; Weiss et al. 1996; Frenois et al. 2002, 2005; Aston-Jones et al. 2004; Diana et al. 1992, 1993, 1995, 1996, 1998, 1999). Finally, all drugs of abuse induce synaptic plasticity at glutamatergic synapses in the VTA in response to a single exposure. The increased excitation of VTA neurons via glutamatergic plasticity is proposed to code the salience of addictive drugs and is linked to behavioral sensitization, a prominent animal model of addiction (Wolf 2003; Robinson and Berridge 1993, 2008; Kauer 2003; Nestler and Aghajanian 1997; De Vries and Shippenberg 2002; Saal et al. 2003; Wolf 1998, 2002; Ungless 2004; Ungless et al. 2001; Mansvelder and McGehee 2000; Faleiro et al. 2004).

Given the central role of DA and VTA DA neurons in the mediation of drugs' effects, a great deal of work has focused on the VTA as the critical site involved in the early neuroadaptations underlying addiction. Many lines of evidence now suggest that the transition to addiction starts in the VTA and involves transient forms of synaptic plasticity that modulate the activity of VTA DA neurons during the first days of acute exposure to drugs and after discontinuing drug treatment. Therefore, current cellular studies continue to use the midbrain slice preparation for examining neuroadaptive changes in the reward circuits with the development of addiction.

Drug-Induced Alterations of GABAergic Function in the VTA

While DA neurons undeniably play an important role in natural and drug-induced reward, GABA interneurons in the VTA also likely mediate some part of reward, either via control of DA neuronal output, or via direct actions of their own. Interestingly, $GABA_A$R- and $GABA_B$ R-containing synapses onto VTA DA neurons initiate from different GABAergic inputs and are functionally distinct (Johnson et al. 1992b; Cameron and Williams 1993; Sugita et al. 1992; Nugent et al. 2009). Whereas GABAergic afferents originating from outside of the VTA, (such as NAc or ventral pallidum) form $GABA_B$R- containing synapses onto VTA DA neurons, GABAergic inputs from the local GABAergic interneurons target $GABA_A$ R- containing synapses (Omelchenko and Sesack 2009; Sugita et al. 1992).

Activation of $GABA_B$ receptors in the VTA inhibits NAc DA release, while activity of $GABA_A$ receptors increases DA release in the NAc. These data suggest that VTA $GABA_A$ receptors are predominantly located on GABAergic interneruons or afferents whose activation disinhibits DA neurons (Xi and Stein 1998, 1999, 2000). This is somewhat surprising, since recordings from DA neurons from *in vitro* preparations reveal the presence of spontaneous large $GABA_A$ responses originating from the VTA (Nugent and Kauer 2008; Sugita et al. 1992; Johnson and North 1992). Moreover, $GABA_A$ antagonists increase DA levels in the NAc, and are self-administered into the VTA. Their rewarding effects are mediated through a DA-dependent mechanism; i.e., through their effects on $GABA_A$ receptors located on VTA DA neurons. All of these effects are characteristic of drugs of abuse (Ikemoto et al. 1997a, b). Opioids and cocaine produce major parts of their reinforcing effects via actions on VTA $GABA_A$ transmission (Johnson and North 1992; Xi and Stein 2002; Liu et al. 2005), as does ethanol, another major abused drug that exerts potent, direct effects on VTA $GABA_A$ synaptic transmission (Melis et al. 2002). Specifically, the effects of opioids on GABAergic transmission in the VTA are prominent. Opioids inhibit release of GABA by activating μ opioid receptors (μOR) localized to VTA GABAergic interneurons and their nerve terminals. Therefore, opioids indirectly excite (disinhibit) DA neurons (Johnson and North 1992; Williams et al. 2001) and consequently the activation of μ opioid receptors in the VTA enhances dopaminergic neurotransmission, increasing both local somatodendritic DA release in the VTA and axonal DA release in the NAc and PFC (Xi and Stein 2002; Klitenick et al. 1992; Spanagel et al. 1992). On the other hand, during chronic morphine withdrawal, DA release is decreased due to increased inhibition of DA neurons in the VTA and the neural substrate of withdrawal seems to engage GABAergic interneurons (Diana et al. 1995; Bonci and Williams 1997). Taken together, these data suggest that $GABA_A$ inhibition in VTA DA neurons may be an important synaptic target for drugs that could modulate VTA DA cell activity and output.

Synaptic Plasticity

Synaptic plasticity is a compelling experimental model to describe the synaptic basis of learning and memory (Bliss and Collingridge 1993). Two forms of synaptic plasticity have been described; LTP and LTD. Based on the synaptic plasticity model, the strength of interneuronal connections (synapses) can be changed as a result of frequent use or of infrequent use. So the frequent use of synapses results in the strengthening of the synapses (LTP) while those synapses that are used infrequently are weakened (LTD) (Wolf 2002). The ability of synapses to exhibit LTD in addition to LTP contributes to the neuronal system flexibility to increase its memory storage capacity (Bear and Abraham 1996). Synaptic plasticity is proposed to subserve a variety of normal brain processes including learning and memory and neuronal development, but similar brain processes appear to be engaged in the pathogenesis of different neuropsychiatric diseases as well as drug addiction.

Drug-Induced Synaptic Plasticity at Excitatory Synapses in the VTA

Historically, the expression of LTP and LTD at excitatory synapses has received more attention than at inhibitory synapses as the cellular substrate of learning. As a result, most of the studies on the neuroplasticity of addiction have focused on synaptic plasticity at excitatory synapses in addiction-related areas of the brain, including the VTA (Kauer and Malenka 2007; Kalivas and O'Brien 2008; Kalivas et al. 2008; Lovinger et al. 2003; Gerdeman et al. 2003). LTP-like and LTD-like modifications of excitatory synapses onto VTA DA neurons can be induced or altered by patterned activity of excitatory afferents or by drugs of abuse (Kauer and Malenka 2007; Bonci and Malenka 1999; Jones and Kauer 1999; Saal et al. 2003;

Ungless et al. 2001; Faleiro et al. 2004; Gutlerner et al. 2002; Jones et al. 2000; Argilli et al. 2008; Borgland et al. 2004; Dong et al. 2004; Thomas et al. 2000, 2001; Thomas and Malenka 2003; Bonci et al. 2008; Borgland et al. 2006; Hahn et al. 2009; Zweifel et al. 2008; Mameli et al. 2007; Bellone and Luscher 2005, 2006; Nugent et al. 2008; Bellone et al. 2008; Engblom et al. 2008; Mameli et al. 2009) suggesting that synaptic plasticity is an ideal candidate for drug-induced reinforcement of addictive behaviors through learning mechanisms (Kauer and Malenka 2007; Hyman and Malenka 2001; Wolf 2002).

The first seminal discovery linking synaptic plasticity to addiction was the demonstration of the induction of cocaine-induced LTP at excitatory synapses onto VTA DA neurons in rats 24 h after a single *in vivo* exposure to cocaine, leading to the proposal that cocaine-induced LTP is responsible for the increased DA cell activity during early stages of behavioral sensitization to cocaine (Ungless et al. 2001). Later, the same group showed that all drugs of abuse, including opioids as well as stress, can produce LTP at the same VTA synapses and that the induction of LTP can be prevented by NMDAR antagonists consistent with the idea that the drug- and stress-induced plasticity are NMDAR-dependent processes (Saal et al. 2003).

An important question is whether it is possible to reverse drug-induced LTP and if so whether LTD (the inverse of LTP) could provide such a mechanism. One indication that this is possible is provided by an elegant study from Luschers' lab which shows that mGluR-dependent LTD in the VTA could reverse the cocaine-induced LTP at glutamatergic synapses through changes in AMPAR distribution. This indicates that triggering the underlying mechanisms of mGluR-LTD could potentially oppose or reverse the LTP associated with behavioral sensitization (Bellone and Luscher 2006; Luscher and Bellone 2008). Moreover, a recent study from the same group showed that local modulation of mGluRs in the VTA DA neurons controls the early and long-lasting forms of cocaine-evoked plasticity in the NAc, presenting the first evidence for a hierarchical link of cocaine-evoked plasticity between the VTA and NAc (Mameli et al. 2009).

Taken together, these studies suggest that glutamatergic synapses onto VTA DA neurons are capable of exhibiting bidirectional plasticity and that drugs of abuse may change the VTA neuronal output through induction and/modulation of different forms of synaptic plasticity at these synapses. Considering the importance of GABAergic inhibition in controlling DA cell excitability, the next critical question is whether GABAergic synapses onto DA neurons are plastic and if so, whether drugs could induce plasticity at these inhibitory synapses that could consequently modulate VTA DA neuronal output.

Drug-Induced GABAergic Synaptic Plasticity in the VTA

While synaptic plasticity at excitatory synapses is well-studied and is correlated with addiction, the role of synaptic plasticity at inhibitory synapses is less well understood. Only a few studies highlight the importance of drug-induced plasticity at inhibitory GABAergic synapses in the VTA in drug addiction (Nugent et al. 2007, 2009; Liu et al. 2005; Melis et al. 2002; Pan et al. 2008a, b). GABAergic inhibition is a key element of essentially every brain circuit, and the control of GABAergic synaptic strength is an important and growing area of interest. However, the study of inhibitory synaptic plasticity in relation to addiction is in its infancy and the molecular nature of drug-induced LTP or LTD at GABAergic synapses is only beginning to be studied.

LTP or LTD of inhibitory GABAergic synapses can modulate the excitability neurons and their firing rate not only through increasing or decreasing $GABA_A$ inhibition respectively, but also through regulation of the induction of synaptic plasticity at adjacent excitatory synapses (i.e., metaplasticity) (Nugent and Kauer 2008). For example, the reduction of GABAergic inhibition by induction of LTD at GABAergic synapses could facilitate the induction of LTP at excitatory synapses. In fact, the cocaine-induced LTD (named as I-LTD) at VTA GABAergic synapses has been demonstrated following *in vivo* repeated exposure to cocaine, and is suggested to facilitate the induction of LTP at VTA excitatory synapses in DA neurons (see below). This mechanism is proposed to underlie the formation of drug-associated memories (Liu et al. 2005; Pan et al. 2008a, b). In the following sections, I will summarize two different forms of LTP found at GABAergic synapses by now in the VTA and an eCB-LTD demonstrated at the same synapses in the VTA, in addition to some other brain areas associated with components of addiction such as the dorsal striatum, amygdala and hippocampus.

LTP of GABAergic Synapses onto VTA DA Neurons

Ethanol-Induced PKA-Dependent LTP

The first demonstration of $GABA_A$R-mediated plasticity in response to drugs of abuse comes from Bonic's group. They showed that by 24 h after a single ethanol treatment, a presynaptic form of LTP at $GABA_A$ synapses onto VTA DA neurons was induced in mice slices. However the underlying mechanism was different from the LTP_{GABA} (Nugent et al. 2007) that we described (see below). Using analyses of paired pulse ratio (PPR) and spontaneous miniature $GABA_A$ IPSCs (mIPSC), they showed that slices from ethanol-treated but not saline-treated mice exhibited ethanol-induced LTP of $GABA_A$ responses apparent as paired pulse depression (PPD) of evoked $GABA_A$ IPSCs and an increase in mIPSCs indicating that the probability of GABA release at these synapses is increased after ethanol (Melis et al. 2002). Interestingly, LTP was prolonged and persisted for a week after treatment. The changes in PPR and mIPSCs in slices from ethanol-treated animals were sensitive to $GABA_B$ receptor blockade suggesting a role for presynaptic $GABA_B$ receptors in mediation of the potentiating effect of ethanol at these synapses. Furthermore, they found that activation of adenylyl cyclase (AC) with forskolin (an AC activator) potentiated evoked $GABA_A$ IPSCs and increased the frequency of mIPSCs in the saline- but not in the ethanol-treated animals. Consistently blocking PKA significantly reduced both frequency of mIPSCs and the amplitude of evoked IPSCs in slices from ethanol-treated mice (Melis et al. 2002), confirming the involvement of cAMP-PKA pathway in ethanol-induced potentiation of GABAergic synapses as shown in Fig. 5.2. The ethanol-induced LTP of GABAergic inhibition is proposed to be critical in ethanol consummatory behavior (Melis et al. 2002). These data provided the first compelling evidence that drugs of abuse can induce neuroadaptive changes by modulating GABAergic plasticity. The next curious question would be whether induction of PKA-dependent LTP at GABAergic synapses is a common theme for all drugs of abuse, similar to what was found about LTP of glutamatergic synapses in the VTA. At this point, at least we know that unlike ethanol, morphine *in vivo* does not affect PKA-dependent LTP at VTA GABAergic synapses (see below). In the coming section, I will discuss another form of a presynaptic LTP at the same GABAergic synapses which in contrast to PKA-dependent LTP is dependent on the NO-cGMP-PKG pathway and is blocked rather than induced by morphine *in vivo*.

Morphine-Blockade of PKG-Dependent LTP

My previous work from Kauer's laboratory provided the first evidence of modulation of LTP at GABAergic synapses by morphine *in vivo* contributing to metaplasticity (Nugent and Kauer 2008; Nugent et al. 2007, 2009). The first critical step to address the effects of drugs of abuse on inhibitory synaptic plasticity is to show whether these synapses are capable of exhibiting any plasticity. Using high frequency stimulation protocol (HFS, the most widely used induction protocol for LTP), we attempted to induce LTP of evoked IPSCs onto DA neurons. We demonstrated that VTA DA neurons but not GABAergic neurons are capable of expressing LTP of GABAergic synapses in response to a pattered stimulation of afferents (LTP_{GABA}). The enhanced inhibition provided by LTP_{GABA} could act as a braking mechanism to counterbalance the increased excitation by LTP at excitatory synapses. Morphine *in vivo* not only induces LTP of excitatory glutamatergic synapses (Saal et al. 2003) but also blocks LTP_{GABA} as we showed recently (Nugent et al. 2007).

To better understand how morphine interacts with LTP_{GABA}, we further explored the molecular mechanisms underlying LTP_{GABA}. As schematically shown in Fig. 5.2, the induction of LTP_{GABA} was postsynaptic and required NMDAR activation and postsynaptic Ca^{2+} influx, however, it appeared to be expressed presynaptically by an increase in GABA release because PPR and the coefficient of variation (CV), the two major indicators of presynaptic plasticity, changed after the induction of LTP_{GABA}. The expression of heterosynaptic LTP_{GABA} is dependent on NO acting as a retrograde messenger. NO generated in a postsynaptic DA neuron in response to activation of NO synthase by intracellular Ca^{2+}, travels retrogradely to activate soluble guanylate cyclase (GC) in neighboring presynaptic GABAergic

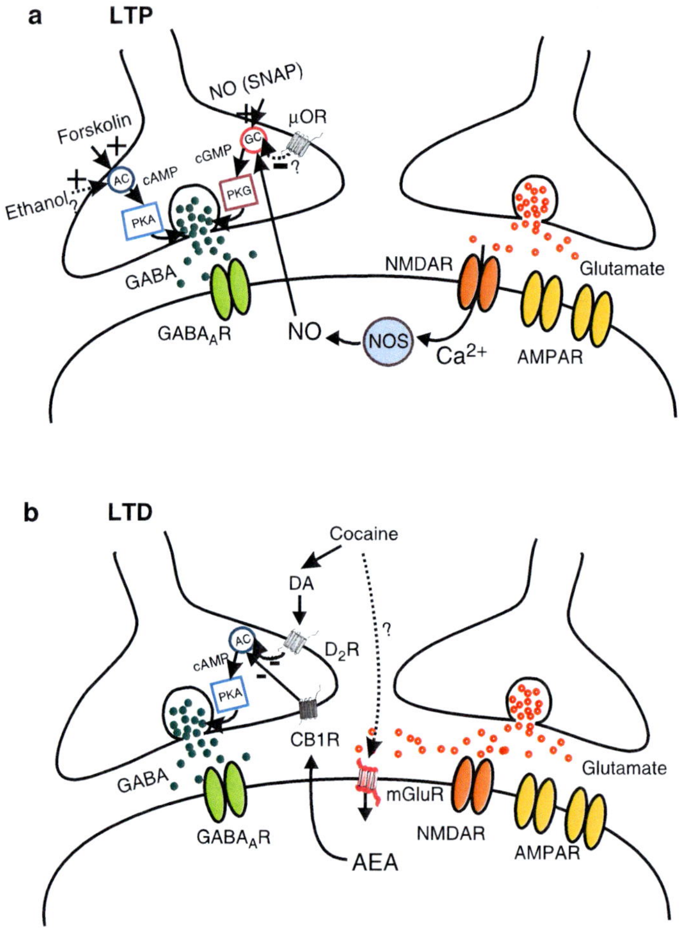

Fig. 5.2 Current models of the signaling molecules involved in GABAergic plasticity of VTA dopamine (DA) neurons and its modulation by drugs of abuse. (**a**) *LTP*: A PKG-dependent presynaptic form of LTP (LTP_{GABA}) is heterosynaptically triggered by postsynaptic NMDAR activation at glutamatergic synapses which consequently engages the NO-cGMP-PKG signaling pathway. Activation of presynaptic cAMP-PKA pathway can also persistently increase GABA release from GABAergic terminals (PKA-dependent LTP). *In vivo* morphine prevents LTP_{GABA} through disruption of the NO signaling pathway probably at the level of GC without affecting the cAMP-PKA pathway. In contrast, *in vivo* ethanol induces PKA-dependent LTP through activation of the cAMP-PKA pathway. (**b**) *LTD*: Activation of D_2Rs facilitates an eCB- mediated LTD of GABAergic synapses onto VTA DA neurons. Activation of group I mGluRs results in production of AEA which in turn activates presynaptic CB1Rs retrogradely. Dopamine signaling through D_2Rs converges on the cAMP-PKA pathway with eCB signaling to cooperatively induce LTD. Repeated exposure to cocaine reduces GABAergic inhibition onto VTA DA neurons through induction of this form of LTD at GABAergic synapses which requires activation of all three receptors; i.e., D_2Rs, mGluRs, and CB1Rs

nerve terminals (Nugent et al. 2007). GC, the main physiological transducer of NO signaling, is aptly named an "NO receptor" because opposite to its widely used name as a soluble enzyme, it is in fact an enzyme-linked receptor which in most cells is indirectly linked to membrane (Garthwaite 2008). Our next experiments confirmed that increased levels of cGMP acting on cGMP-dependent protein kinase (PKG) in presynaptic terminals promoted long lasting potentiation of GABA release at these synapses (Nugent et al. 2009). Synaptic plasticity can be stimulated in various parts of the brain by NO acting on the cGMP- PKG pathway (Lu and Hawkins 2002; Lu et al. 1999; Monfort et al. 2002, 2004; Chien et al. 2003; Huang et al. 2003; Liu et al. 2003; Zhuo et al. 1994). We also found that the potentiating effects of NO and cGMP on GABAergic synapses were blocked by a PKG inhibitor, KT5823, confirming that PKG is downstream to NO-cGMP. Furthermore, we demonstrated that blockade of PKG inhibited the induction of LTP_{GABA} while the maintenance of LTP_{GABA} was unaffected (Nugent et al. 2009) suggesting that the transient activation of NO signaling pathway is enough to trigger long-term plasticity but not necessary to maintain the plasticity. It seems that the induction of LTP_{GABA} requires an additional molecular mechanism at the presynaptic terminal to translate transient PKG signaling into a persistent potentiation that is PKG-independent. In our experiments we have only tried the effect of PKG blocker 10 min after HFS. It will be interesting to see for how long activation of PKG is needed to induce LTP_{GABA}.

The cAMP-PKA pathway coexists with the cGMP-PKG pathway in the VTA GABAergic terminals and the activation of this pathway also enhances GABA release from these terminals persistently (Nugent et al. 2009; Bonci and Williams 1997). Given that the activation of the cAMP pathway is involved in many forms of synaptic plasticity and can also facilitate the transmitter release at many synapses and that ethanol can induce a PKA-dependent LTP at these synapses, we tested the effects of drugs targeting this pathway on LTP_{GABA}. We found that activation of AC by forskolin or PKA by Sp-cAMPS, the cAMP agonist, potentiated GABAergic synapses onto DA neurons and the potentiation was accompanied by a decrease in the PPR, suggesting an enhanced GABA release at these synapses. Furthermore, the prior potentiation of synapses by forskolin and Sp-cAMPS occluded HFS-induced LTP_{GABA} indicating that the mechanisms underlying LTP_{GABA} and the cAMP-PKA dependent potentiation of GABAergic synapses are shared. However, when we attempted to induce LTP_{GABA} in the presence of PKA blocker, Rp-cAMPS, it did not affect the induction or the maintenance of LTP_{GABA} demonstrating that the NO-cGMP-PKG pathway, but not the cAMP-PKA pathway, is critical for inducing LTP_{GABA} (Nugent et al. 2009). Using a similar approach, we conducted a series of saturation-occlusion experiments to test the idea whether PKG and PKA converge onto a common mechanism to maintain the increased GABA release at these synapses. We found that prior potentiation of evoked $GABA_A$ IPSCs by an NO donor (S-nitroso-N-acetylpenicillamine, SNAP) occludes further potentiation of $GABA_A$ responses with forskolin suggesting that PKG and PKA may converge on a common downstream signal that is necessary for LTP_{GABA}. It will be very interesting to identify by which mechanisms the transient activation of PKA and PKG translates to long-term changes of the presynaptic GABA release (i.e., maintenance of LTP_{GABA}). It seems that phosphorylation of a common substrate for both PKA and PKG which is also critical in the presynaptic plasticity would be the answer. Identification of this convergent signal merits further investigation.

As mentioned above raising the levels of either cGMP (by an NO donor) or cAMP (by forskolin) increased presynaptic GABA release persistently and mimicked LTP_{GABA} (see Fig. 5.2). Next we explored the molecular mechanisms by which morphine could block LTP_{GABA}. We examined the effects of the NO donor (SNAP), the cGMP analogue (pCPT-cGMP, 8-(p-chlorophenylthio)-cGMP) and the AC activator (forskolin) on slices prepared from saline and morphine treated rats. The results showed that GABAergic synapses from slices prepared from saline- and morphine-treated rats 24 h after *in vivo* morphine were equally potentiated in response to bath application of forskolin, suggesting that contrary to ethanol, the cAMP-PKA pathway in VTA GABAergic terminals was unaffected by morphine (Nugent et al. 2009). In contrast, while GABAergic synapses from saline-treated rats were potentiated in response to NO released by SNAP; synapses from morphine-treated rats were unaffected, suggesting that morphine *in vivo* changed the ability of synapses to respond to NO. In an attempt to determine the site of disruption of the NO signaling pathway by morphine, morphine-treated slices were exposed to pCPT-cGMP, the cGMP agonist. Interestingly, we found that the synapses from morphine-treated rats were potentiated in response to bath application of pCPT-cGMP, indicating that the site of interaction is at the level of GC (Nugent et al. 2007). The selectivity of acute morphine in alteration of the NO-cGMP signaling rather than the cAMP pathway highlights the fact that the two signaling pathways can be modulated independently. It is still possible that unlike acute exposure, chronic exposure to morphine involves alterations in both signaling pathways. To date, there is no evidence suggesting a direct interaction between opioid receptors and GC, however it is possible that through other membrane associated proteins interacting with this enzyme-linked receptor, opioid receptors may be coupled indirectly to GC and modulate cGMP production. Blockade of LTP_{GABA} 24 h post-morphine *in vivo* suggests a long-lasting effect of morphine on the VTA circuitry that may be mediated and consolidated through morphine's effect on gene transcription when morphine is no longer present. Our data also suggest that drugs targeting cAMP or cGMP pathways in the VTA GABAergic synapses could be efficacious in reversing morphine-induced maladaptive changes at these synapses. Given that the VTA plays a key role in the aversive aspect of opiate withdrawal, and that the negative affective state of withdrawal promotes drug-seeking behavior (Frenois et al. 2002; Frenois et al. 2005), it is tempting to speculate that LTP_{GABA}, by increasing inhibition of DA neurons under pathological conditions, could result in negative affective states such as the dysphoria or anhedonia associated with decreased DA levels in midbrain. It will be critical to examine the time course of morphine-blockade of LTP_{GABA}, and show whether other drugs of abuse, as well as stress, are capable of modulating NO-dependent LTP_{GABA}. Unlike the ethanol-induce changes in PPR in slices from ethanol-treated mice, the values of baseline PPRs of slices from morphine-treated rats were not different from those from saline-treated rats, suggesting that contrary to the ethanol's potentiating effect on presynaptic GABA release, morphine blocks rather than induces LTP of GABAergic synapses. It is reasonable to speculate that ethanol might modulate NO-mediated LTP_{GABA} in a similar manner to PKA-dependent LTP. Whether the downstream convergent mechanism for PKA and PKG would be a target for morphine, ethanol and other drugs of abuse is an idea worth pursuing and will require further study.

Cocaine-Induced eCB-Mediated LTD

In a pioneering study 7 years ago, Lovinger's group presented the first demonstration of the involvement of eCBs as retrograde signaling molecules in long-term synaptic plasticity in the dorsal striatum (eCB-LTD) (Gerdeman et al. 2002). Since then, others have reported eCB-LTD at both excitatory glutamatergic (Sjostrom et al. 2003, 2004; Mato et al. 2008; Lafourcade et al. 2007; Yasuda et al. 2008) and inhibitory GABAergic synapses (Chevaleyre et al. 2007; Pan et al. 2008a; Chevaleyre and Castillo 2003; Marsicano et al. 2002; Adermark et al. 2009) in several brain structures. While the topic of eCB-mediated LTD is a garnering interest among neuroscientists, the existence of inhibitory eCB-LTD or its possible modulation by drugs of abuse has not been tested extensively in some of the structures that are critically associated with components of addiction including the NAc, PFC, amygdala, and hippocampus. For example, while the medium spiny neurons (MSNs) of dorsal striatum exhibit eCB-LTD of GABAergic inhibition (Adermark et al. 2009), it is not known whether this plasticity can be expressed in MSNs of NAc. Given the similarities between the ventral and dorsal striatal MSNs and that mesoaccumbal and nigrostriatal dopamine systems are both proposed to participate in reward and addiction (Wise and Roles for nigrostriatal-not just mesocorticolimbic-dopamine in reward and addiction. Trends Neurosci 2009), it is reasonable to speculate that eCB-LTD of GABAergic synapses exist in NAc. This presents a potential valid target for drugs of abuse.

Because of the comprehensive recent reviews on eCB-mediated LTD, I will briefly summarize the underlying mechanisms of its induction and expression below and then mention one example of this form of plasticity that is recently found in the VTA. An eCB-LTD can be dependent or independent of increases in postsynaptic Ca^{2+}, therefore in some instances eCB-LTD requires activation of NMDARs or voltage-gated calcium channels or Ca^{2+} release from endoplasmic reticulum (Lovinger 2008; Heifets and Castillo 2009). One of the most common triggers of production of eCBs in postsynaptic cell is activation of group I mGluRs by released glutamate from neighboring glutamatergic terminals. Activation of mGluRs initiates a cascade of events including activation of phospholipase C (PLC) and diacylglycerol lipase (DGL) to ultimately generate the eCB, 2-arachidonyl glycerol (2-AG) (Heifets and Castillo 2009; Chevaleyre and Castillo 2003). Independently, activation of postsynaptic D_2Rs can also result in formation of another eCB, anandamide (AEA), which is important in corticostriatal and amygdalar eCB-LTD (Ronesi and Lovinger 2005; Kreitzer and Malenka 2005, 2007, 2008; Azad et al. 2004).

The generated eCBs in response to these postsynaptic triggers act as a retrograde messenger, inducing eCB-LTD through presynaptic CB1 receptors (CB_1Rs) that are negatively coupled to the cAMP-PKA pathway. By reducing PKA activity, which in some cases results in dephosphorylation of the active zone protein RIM_1alpha downstream to CB_1Rs, the release machinery in presynaptic terminal will be affected permanently (Chevaleyre et al. 2007).

An eCB-mediated LTD of inhibitory GABAergic synapses expressed in VTA DA neurons can be modulated by repeated *in vivo* exposure to cocaine as shown recently (Pan et al. 2008a, b) (also see Fig. 5.2). In this study from Liu's laboratory, eCB-LTD is triggered using moderate 10 Hz synaptic stimulation combined with cocaine application ("cocaine-induced I-LTD"). The downside of the complexity of I-LTD protocol is the likelihood of the simultaneous trigger of several forms of pre- and post synaptic plasticity by *in vitro* cocaine application because of cocaine's inhibitory effect on uptake of other neurotransmitters besides dopamine. Therefore, one alternative is to achieve LTD induction using electrical stimulation protocols. The cocaine-induced I-LTD was independent of postsynaptic Ca^{2+} and was occluded after repeated rather than single exposure to cocaine, suggesting that chronic *in vivo* cocaine in fact induces I-LTD to reduce GABAergic inhibition in VTA DA neurons.

The observation that PPR and CV measurements, as well as TTX-sensitive mIPSCs were all unaffected after the induction of cocaine-induced I-LTD is suggestive of a postsynaptic form of I-LTD which is in contrast to other forms of eCB-LTD with a presynaptic nature of expression (Chevaleyre and Castillo 2003). In an attempt to distinguish the site of I-LTD expression, these authors analyzed strontium (Sr^+)-induced asynchronous quantal events (Sr^+ mIPSCs). Cocaine-induced I-LTD was in fact associated with both a decrease in the frequency and amplitude of these quantal events suggesting that both pre- and post-synaptic mechanisms were participating in the expression of cocaine-induced I-LTD. Using blockers and other pharmacological and genetic manipulations of D_2 dopamine receptors (D_2Rs), CB_1Rs, and group I mGluRs, they further characterized I-LTD and showed that the induction of I-LTD requires activation of all three receptors. Additional disruption of endocannabinoid synthesis (in this case 2AG) by blockers of PLC and DGL prevented I-LTD, suggesting that the production of 2AG is coupled to activation of mGluRs for induction of I-LTD. Interestingly, *in vivo* pretreatment of animals with antagonists of all three receptors blocked the *in vivo* cocaine-induced I-LTD apparent as an increase in the amplitude of maximal evoked IPSCs (Pan et al. 2008a, b).

Further analysis of I-LTD through a series of exhaustive experiments by this group revealed that D_2R activation facilitates mGluR-2AG-CB_1R-mediated I-LTD through inhibition of the presynaptic cAMP-PKA pathway rather than enhancing the eCB, AEA, postsynaptically. The latter is proposed by the prevailing model of striatal plasticity supporting a cooperative production of eCBs by D_2Rs and mGluRs to induce eCB-LTD (Kreitzer and Malenka 2005, 2007). However, based on this novel model (see Fig. 5.2); the presynaptic cAMP-PKA pathway is the convergent downstream target for both D_2Rs and eCB signaling to induce I-LTD.

This study convincingly supports a role for an inhibitory synaptic plasticity mediated by both eCB and dopamine signals in cocaine-induced modification of

VTA circuitry. Existence of different forms of LTD of inhibitory synapses in the VTA is very likely. In fact, our recent preliminary data suggest the presence of a postsynaptic LTD of GABAergic synapses in the VTA in response to an electrical stimulation paradigm which seems to be sensitive to a single exposure to morphine *in vivo* (unpublished data). Whether morphine, cocaine and other drugs are capable of modulating several or all forms of these plastic changes in the VTA are unclear and merit future investigation.

Conclusions

Here I briefly reviewed three forms of synaptic plasticity recently described at GABAergic synapses of the VTA and their possible link to addictive properties of some drugs of abuse. The field of inhibitory synaptic plasticity is young and growing fast and excitingly its critical role in different brain processes such as fear extinction, sensory deprivation and now drug addiction are only now beginning to be uncovered. Is it possible that the amygdalar eCB-LTD, important in erasing fear memories, also could play a role in extinction of withdrawal-associated negative memories? Is it possible to prevent or reverse the process of induction and consolidation of addictive memories? Induction of LTP and LTD with opposite alterations of synaptic efficacy presents the possibility of the interaction of their underlying mechanisms. This level of interaction and reversibility offers counterbalancing mechanisms to reset synaptic efficacy or reverse a drug-induced aberrant form of plasticity. Therefore, it is of great interest to continue probing the different forms of synaptic plasticity at excitatory as well as inhibitory synapses in not only the VTA but other brain areas important in addiction. There is hope that uncovering the molecular mechanisms underlying different forms of synaptic plasticity in addiction-associated structures will provide us with novel molecular drug targets to prevent the pathological learning of addiction and ultimately eradicate addiction-related memories.

Acknowledgements The opinions and assertions contained herein are the private opinions of the author and are not to be construed as official or reflecting the views of the Uniformed Services University of the Health Sciences or the Department of Defense or the Government of the United States. This work was supported by an R0 75OU grant from the Uniformed Services University (USUHS), and I also acknowledge past support (5F32 DA021973-02) from the National Institute of Drug Abuse. Thanks to Drs. Brian Cox, David Lovinger, Julie Kauer, and Suzanne Bausch for their helpful and constructive discussions for the present chapter.

References

Adermark, L., G. Talani, and D.M. Lovinger, Endocannabinoid-dependent plasticity at GABAergic and glutamatergic synapses in the striatum is regulated by synaptic activity. Eur J Neurosci, 2009. **29**(1): p. 32–41.

Argilli, E., et al., Mechanism and time course of cocaine-induced long-term potentiation in the ventral tegmental area. J Neurosci, 2008. **28**(37): p. 9092–100.
Aston-Jones, G. and G.C. Harris, Brain substrates for increased drug seeking during protracted withdrawal. Neuropharmacology, 2004. **47 Suppl 1**: p. 167–79.
Azad, S.C., et al., Circuitry for associative plasticity in the amygdala involves endocannabinoid signaling. J Neurosci, 2004. **24**(44): p. 9953–61.
Bear, M.F. and W.C. Abraham, Long-term depression in hippocampus. Annu Rev Neurosci, 1996. **19**: p. 437–62.
Bellone, C. and C. Luscher, mGluRs induce a long-term depression in the ventral tegmental area that involves a switch of the subunit composition of AMPA receptors. Eur J Neurosci, 2005. **21**(5): p. 1280–8.
Bellone, C. and C. Luscher, Cocaine triggered AMPA receptor redistribution is reversed in vivo by mGluR-dependent long-term depression. Nat Neurosci, 2006. **9**(5): p. 636–41.
Bellone, C., C. Luscher, and M. Mameli, Mechanisms of synaptic depression triggered by metabotropic glutamate receptors. Cell Mol Life Sci, 2008. **65**(18): p. 2913–23.
Betz, C., et al., Could a common biochemical mechanism underlie addictions? J Clin Pharm Ther, 2000. **25**(1): p. 11–20.
Bliss, T.V. and G.L. Collingridge, A synaptic model of memory: long-term potentiation in the hippocampus. Nature, 1993. **361**(6407): p. 31–9.
Bonci, A. and R.C. Malenka, Properties and plasticity of excitatory synapses on dopaminergic and GABAergic cells in the ventral tegmental area. J Neurosci, 1999. **19**(10): p. 3723–30.
Bonci, A. and J.T. Williams, A common mechanism mediates long-term changes in synaptic transmission after chronic cocaine and morphine. Neuron, 1996. **16**(3): p. 631–9.
Borgland, S.L., et al., Orexin A in the VTA is critical for the induction of synaptic plasticity and behavioral sensitization to cocaine. Neuron, 2006. **49**(4): p. 589–601.
Borgland, S.L., R.C. Malenka, and A. Bonci, Acute and chronic cocaine-induced potentiation of synaptic strength in the ventral tegmental area: electrophysiological and behavioral correlates in individual rats. J Neurosci, 2004. **24**(34): p. 7482–90.
Bonci, A. and S. Borgland, Role of orexin/hypocretin and CRF in the formation of drug-dependent synaptic plasticity in the mesolimbic system. Neuropharmacology, 2008.
Bonci, A. and J.T. Williams, Increased probability of GABA release during withdrawal from morphine. J Neurosci, 1997. **17**(2): p. 796–803.
Cameron, D.L. and J.T. Williams, Dopamine D1 receptors facilitate transmitter release. Nature, 1993. **366**(6453): p. 344–7.
Carr, D.B. and S.R. Sesack, Projections from the rat prefrontal cortex to the ventral tegmental area: target specificity in the synaptic associations with mesoaccumbens and mesocortical neurons. J Neurosci, 2000. **20**(10): p. 3864–73.
Chergui, K., et al., Tonic activation of NMDA receptors causes spontaneous burst discharge of rat midbrain dopamine neurons in vivo. Eur J Neurosci, 1993. **5**(2): p. 137–44.
Chevaleyre, V. and P.E. Castillo, Heterosynaptic LTD of hippocampal GABAergic synapses: a novel role of endocannabinoids in regulating excitability. Neuron, 2003. **38**(3): p. 461–72.
Chevaleyre, V., K.A. Takahashi, and P.E. Castillo, Endocannabinoid-mediated synaptic plasticity in the CNS. Annu Rev Neurosci, 2006. **29**: p. 37–76.
Chevaleyre, V., et al., Endocannabinoid-mediated long-term plasticity requires cAMP/PKA signaling and RIM1alpha. Neuron, 2007. **54**(5): p. 801–12.
Chien, W.L., et al., Enhancement of long-term potentiation by a potent nitric oxide-guanylyl cyclase activator, 3-(5-hydroxymethyl-2-furyl)-1-benzyl-indazole. Mol Pharmacol, 2003. **63**(6): p. 1322–8.
Churchill, L., R.P. Dilts, and P.W. Kalivas, Autoradiographic localization of gamma-aminobutyric acidA receptors within the ventral tegmental area. Neurochem Res, 1992. **17**(1): p. 101–6.
Christie, M.J., et al., Excitotoxin lesions suggest an aspartatergic projection from rat medial prefrontal cortex to ventral tegmental area. Brain Res, 1985. **333**(1): p. 169–72.
De Vries, T.J. and T.S. Shippenberg, Neural systems underlying opiate addiction. J Neurosci, 2002. **22**(9): p. 3321–5.

Diana, M., et al., Profound decrement of mesolimbic dopaminergic neuronal activity during ethanol withdrawal syndrome in rats: electrophysiological and biochemical evidence. Proc Natl Acad Sci U S A, 1993. **90**(17): p. 7966–9.
Diana, M., et al., Mesolimbic dopaminergic reduction outlasts ethanol withdrawal syndrome: evidence of protracted abstinence. Neuroscience, 1996. **71**(2): p. 411–5.
Diana, M., et al., Marked decrease of A10 dopamine neuronal firing during ethanol withdrawal syndrome in rats. Eur J Pharmacol, 1992. **221**(2–3): p. 403–4.
Diana, M., et al., Profound decrease of mesolimbic dopaminergic neuronal activity in morphine withdrawn rats. J Pharmacol Exp Ther, 1995. **272**(2): p. 781–5.
Diana, M., et al., Lasting reduction in mesolimbic dopamine neuronal activity after morphine withdrawal. Eur J Neurosci, 1999. **11**(3): p. 1037–41.
Diana, M., et al., Mesolimbic dopaminergic decline after cannabinoid withdrawal. Proc Natl Acad Sci U S A, 1998. **95**(17): p. 10269–73.
Di Chiara, G. and A. Imperato, Drugs abused by humans preferentially increase synaptic dopamine concentrations in the mesolimbic system of freely moving rats. Proc Natl Acad Sci U S A, 1988. **85**(14): p. 5274–8.
Dong, Y., et al., Cocaine-induced potentiation of synaptic strength in dopamine neurons: behavioral correlates in GluRA(–/–) mice. Proc Natl Acad Sci U S A, 2004. **101**(39): p. 14282–7.
Engblom, D., et al., Glutamate receptors on dopamine neurons control the persistence of cocaine seeking. Neuron, 2008. **59**(3): p. 497–508.
Faleiro, L.J., S. Jones, and J.A. Kauer, Rapid synaptic plasticity of glutamatergic synapses on dopamine neurons in the ventral tegmental area in response to acute amphetamine injection. Neuropsychopharmacology, 2004. **29**(12): p. 2115–25.
Fields, H.L., et al., Ventral Tegmental Area Neurons in Learned Appetitive Behavior and Positive Reinforcement. Annu Rev Neurosci, 2007.
Fiorillo, C.D. and J.T. Williams, Glutamate mediates an inhibitory postsynaptic potential in dopamine neurons. Nature, 1998. **394**(6688): p. 78–82.
Frenois, F., et al., A specific limbic circuit underlies opiate withdrawal memories. J Neurosci, 2005. **25**(6): p. 1366–74.
Frenois, F., et al., Neural correlates of the motivational and somatic components of naloxone-precipitated morphine withdrawal. Eur J Neurosci, 2002. **16**(7): p. 1377–89.
Garthwaite, J., Concepts of neural nitric oxide-mediated transmission. Eur J Neurosci, 2008. **27**(11): p. 2783–802.
Garzon, M., et al., Cholinergic axon terminals in the ventral tegmental area target a subpopulation of neurons expressing low levels of the dopamine transporter. J Comp Neurol, 1999. **410**(2): p. 197–210.
Georges, F. and G. Aston-Jones, Potent regulation of midbrain dopamine neurons by the bed nucleus of the stria terminalis. J Neurosci, 2001. **21**(16): p. RC160.
Gerdeman, G.L. and D.M. Lovinger, Emerging roles for endocannabinoids in long-term synaptic plasticity. Br J Pharmacol, 2003. **140**(5): p. 781–9.
Gerdeman, G.L., et al., It could be habit forming: drugs of abuse and striatal synaptic plasticity. Trends Neurosci, 2003. **26**(4): p. 184–92.
Gerdeman, G.L., J. Ronesi, and D.M. Lovinger, Postsynaptic endocannabinoid release is critical to long-term depression in the striatum. Nat Neurosci, 2002. **5**(5): p. 446–51.
Gutlerner, J.L., et al., Novel protein kinase A-dependent long-term depression of excitatory synapses. Neuron, 2002. **36**(5): p. 921–31.
Hahn, J., F.W. Hopf, and A. Bonci, Chronic cocaine enhances corticotropin-releasing factor-dependent potentiation of excitatory transmission in ventral tegmental area dopamine neurons. J Neurosci, 2009. **29**(20): p. 6535–44.
Heifets, B.D. and P.E. Castillo, Endocannabinoid signaling and long-term synaptic plasticity. Annu Rev Physiol, 2009. **71**: p. 283–306.
Huang, C.C., S.H. Chan, and K.S. Hsu, cGMP/protein kinase G-dependent potentiation of glutamatergic transmission induced by nitric oxide in immature rat rostral ventrolateral medulla neurons in vitro. Mol Pharmacol, 2003. **64**(2): p. 521–32.

Hyman, S.E. and R.C. Malenka, Addiction and the brain: the neurobiology of compulsion and its persistence. Nat Rev Neurosci, 2001. **2**(10): p. 695–703.

Ikemoto, S., R.R. Kohl, and W.J. McBride, GABA(A) receptor blockade in the anterior ventral tegmental area increases extracellular levels of dopamine in the nucleus accumbens of rats. J Neurochem, 1997. **69**(1): p. 137–43.

Ikemoto, S., J.M. Murphy, and W.J. McBride, Self-infusion of GABA(A) antagonists directly into the ventral tegmental area and adjacent regions. Behav Neurosci, 1997. **111**(2): p. 369–80.

Jones, S., J.L. Kornblum, and J.A. Kauer, Amphetamine blocks long-term synaptic depression in the ventral tegmental area. J Neurosci, 2000. **20**(15): p. 5575–80.

Jones, S. and J.A. Kauer, Amphetamine depresses excitatory synaptic transmission via serotonin receptors in the ventral tegmental area. J Neurosci, 1999. **19**(22): p. 9780–7.

Johnson, S.W., N.B. Mercuri, and R.A. North, 5-hydroxytryptamine1B receptors block the GABAB synaptic potential in rat dopamine neurons. J Neurosci, 1992. **12**(5): p. 2000–6.

Johnson, S.W., V. Seutin, and R.A. North, Burst firing in dopamine neurons induced by N-methyl-D-aspartate: role of electrogenic sodium pump. Science, 1992. **258**(5082): p. 665–7.

Johnson, S.W. and R.A. North, Opioids excite dopamine neurons by hyperpolarization of local interneurons. J Neurosci, 1992. **12**(2): p. 483–8.

Kalivas, P.W. and C. O'Brien, Drug addiction as a pathology of staged neuroplasticity. Neuropsychopharmacology, 2008. **33**(1): p. 166–80.

Kalivas, P.W., et al., Glutamate transmission in addiction. Neuropharmacology, 2008.

Kauer, J.A., Addictive drugs and stress trigger a common change at VTA synapses. Neuron, 2003. **37**(4): p. 549–50.

Kauer, J.A. and R.C. Malenka, Synaptic plasticity and addiction. Nat Rev Neurosci, 2007. **8**(11): p. 844–58.Volkow, N. and T.K. Li, The neuroscience of addiction. Nat Neurosci, 2005. **8**(11): p. 1429–30.

Klitenick, M.A., P. DeWitte, and P.W. Kalivas, Regulation of somatodendritic dopamine release in the ventral tegmental area by opioids and GABA: an in vivo microdialysis study. J Neurosci, 1992. **12**(7): p. 2623–32.

Koob, G.F., T.L. Wall, and F.E. Bloom, Nucleus accumbens as a substrate for the aversive stimulus effects of opiate withdrawal. Psychopharmacology (Berl), 1989. **98**(4): p. 530–4.

Koob, G.F., G. Kenneth Lloyd, and B.J. Mason, Development of pharmacotherapies for drug addiction: a Rosetta stone approach. Nat Rev Drug Discov, 2009. **8**(6): p. 500–15.

Koob, G.F., A role for brain stress systems in addiction. Neuron, 2008. **59**(1): p. 11–34.

Koob, G.F., Neural mechanisms of drug reinforcement. Ann N Y Acad Sci, 1992. **654**: p. 171–91.

Koob, G.F. and M. Le Moal, Drug abuse: hedonic homeostatic dysregulation. Science, 1997. **278**(5335): p. 52–8.

Koob, G.F. and M. Le Moal, Plasticity of reward neurocircuitry and the 'dark side' of drug addiction. Nat Neurosci, 2005. **8**(11): p. 1442–4.

Koob, G.F. and N.D. Volkow, Neurocircuitry of Addiction. Neuropsychopharmacology, 2009.

Koob, G.F. and M. Le Moal, Drug addiction, dysregulation of reward, and allostasis. Neuropsychopharmacology, 2001. **24**(2): p. 97–129.

Kreitzer, A.C. and R.C. Malenka, Dopamine modulation of state-dependent endocannabinoid release and long-term depression in the striatum. J Neurosci, 2005. **25**(45): p. 10537–45.

Kreitzer, A.C. and R.C. Malenka, Striatal plasticity and basal ganglia circuit function. Neuron, 2008. **60**(4): p. 543–54.

Kreitzer, A.C. and R.C. Malenka, Endocannabinoid-mediated rescue of striatal LTD and motor deficits in Parkinson's disease models. Nature, 2007. **445**(7128): p. 643–7.

Lafourcade, M., et al., Molecular components and functions of the endocannabinoid system in mouse prefrontal cortex. PLoS ONE, 2007. **2**(1): p. e709.

Liu, Z.H., R. Shin, and S. Ikemoto, Dual role of medial A10 dopamine neurons in affective encoding. Neuropsychopharmacology, 2008. **33**(12): p. 3010–20.

Liu, Q.S., L. Pu, and M.M. Poo, Repeated cocaine exposure in vivo facilitates LTP induction in midbrain dopamine neurons. Nature, 2005. **437**(7061): p. 1027–31.

Liu, S., Y. Rao, and N. Daw, Roles of protein kinase A and protein kinase G in synaptic plasticity in the visual cortex. Cereb Cortex, 2003. **13**(8): p. 864–9.
Lovinger, D.M., Presynaptic modulation by endocannabinoids. Handb Exp Pharmacol, 2008(184): p. 435–77.
Lovinger, D.M., J.G. Partridge, and K.C. Tang, Plastic control of striatal glutamatergic transmission by ensemble actions of several neurotransmitters and targets for drugs of abuse. Ann N Y Acad Sci, 2003. **1003**: p. 226–40.
Luscher, C. and C. Bellone, Cocaine-evoked synaptic plasticity: a key to addiction? Nat Neurosci, 2008. **11**(7): p. 737–8.
Lu, Y.F. and R.D. Hawkins, Ryanodine receptors contribute to cGMP-induced late-phase LTP and CREB phosphorylation in the hippocampus. J Neurophysiol, 2002. **88**(3): p. 1270–8.
Lu, Y.F., E.R. Kandel, and R.D. Hawkins, Nitric oxide signaling contributes to late-phase LTP and CREB phosphorylation in the hippocampus. J Neurosci, 1999. **19**(23): p. 10250–61.
Mameli, M., et al., Cocaine-evoked synaptic plasticity: persistence in the VTA triggers adaptations in the NAc. Nat Neurosci, 2009.
Mameli, M., et al., Rapid synthesis and synaptic insertion of GluR2 for mGluR-LTD in the ventral tegmental area. Science, 2007. **317**(5837): p. 530–3.
Mansvelder, H.D. and D.S. McGehee, Long-term potentiation of excitatory inputs to brain reward areas by nicotine. Neuron, 2000. **27**(2): p. 349–57.
Marsicano, G., et al., The endogenous cannabinoid system controls extinction of aversive memories. Nature, 2002. **418**(6897): p. 530–4.
Mato, S., et al., Role of the cyclic-AMP/PKA cascade and of P/Q-type Ca++ channels in endocannabinoid-mediated long-term depression in the nucleus accumbens. Neuropharmacology, 2008. **54**(1): p. 87–94.
McBain, C.J. and J.A. Kauer, Presynaptic plasticity: targeted control of inhibitory networks. Curr Opin Neurobiol, 2009.
Melis, M., et al., Long-lasting potentiation of GABAergic synapses in dopamine neurons after a single in vivo ethanol exposure. J Neurosci, 2002. **22**(6): p. 2074–82.
Monfort, P., et al., Long-term potentiation in hippocampus involves sequential activation of soluble guanylate cyclase, cGMP-dependent protein kinase, and cGMP-degrading phosphodiesterase. J Neurosci, 2002. **22**(23): p. 10116–22.
Monfort, P., et al., Sequential activation of soluble guanylate cyclase, protein kinase G and cGMP-degrading phosphodiesterase is necessary for proper induction of long-term potentiation in CA1 of hippocampus. Alterations in hyperammonemia. Neurochem Int, 2004. **45**(6): p. 895–901.
Murase, S., et al., Prefrontal cortex regulates burst firing and transmitter release in rat mesolimbic dopamine neurons studied in vivo. Neurosci Lett, 1993. **157**(1): p. 53–6.
Nestler, E.J. and G.K. Aghajanian, Molecular and cellular basis of addiction. Science, 1997. **278**(5335): p. 58–63.
Nugent, F.S., J.L. Niehaus, and J.A. Kauer, PKG and PKA Signaling in LTP at GABAergic Synapses. Neuropsychopharmacology, 2009.
Nugent, F.S., et al., High-frequency afferent stimulation induces long-term potentiation of field potentials in the ventral tegmental area. Neuropsychopharmacology, 2008. **33**(7): p. 1704–12.
Nugent, F.S., E.C. Penick, and J.A. Kauer, Opioids block long-term potentiation of inhibitory synapses. Nature, 2007. **446**(7139): p. 1086–90.
Nugent, F.S. and J.A. Kauer, LTP of GABAergic synapses in the ventral tegmental area and beyond. J Physiol, 2008. **586**(6): p. 1487–93.
Omelchenko, N. and S.R. Sesack, Ultrastructural analysis of local collaterals of rat ventral tegmental area neurons: GABA phenotype and synapses onto dopamine and GABA cells. Synapse, 2009. **63**(10): p. 895–906.
Pan, B., C.J. Hillard, and Q.S. Liu, Endocannabinoid signaling mediates cocaine-induced inhibitory synaptic plasticity in midbrain dopamine neurons. J Neurosci, 2008. **28**(6): p. 1385–97.
Pan, B., C.J. Hillard, and Q.S. Liu, D2 dopamine receptor activation facilitates endocannabinoid-mediated long-term synaptic depression of GABAergic synaptic transmission in midbrain

dopamine neurons via cAMP-protein kinase A signaling. J Neurosci, 2008. **28**(52): p. 14018–30.

Robinson, T.E. and K.C. Berridge, The neural basis of drug craving: an incentive-sensitization theory of addiction. Brain Res Brain Res Rev, 1993. **18**(3): p. 247–91.

Robinson, T.E. and K.C. Berridge, Review. The incentive sensitization theory of addiction: some current issues. Philos Trans R Soc Lond B Biol Sci, 2008. **363**(1507): p. 3137–46.

Ronesi, J. and D.M. Lovinger, Induction of striatal long-term synaptic depression by moderate frequency activation of cortical afferents in rat. J Physiol, 2005. **562**(Pt 1): p. 245–56.

Saal, D., et al., Drugs of abuse and stress trigger a common synaptic adaptation in dopamine neurons. Neuron, 2003. **37**(4): p. 577–82.

Schultz, W., Dopamine neurons and their role in reward mechanisms. Curr Opin Neurobiol, 1997. **7**(2): p. 191–7.

Sesack, S.R. and V.M. Pickel, Prefrontal cortical efferents in the rat synapse on unlabeled neuronal targets of catecholamine terminals in the nucleus accumbens septi and on dopamine neurons in the ventral tegmental area. J Comp Neurol, 1992. **320**(2): p. 145–60.

Semba, K. and H.C. Fibiger, Afferent connections of the laterodorsal and the pedunculopontine tegmental nuclei in the rat: a retro- and antero-grade transport and immunohistochemical study. J Comp Neurol, 1992. **323**(3): p. 387–410.

Sjostrom, P.J., G.G. Turrigiano, and S.B. Nelson, Neocortical LTD via coincident activation of presynaptic NMDA and cannabinoid receptors. Neuron, 2003. **39**(4): p. 641–54.

Sjostrom, P.J., G.G. Turrigiano, and S.B. Nelson, Endocannabinoid-dependent neocortical layer-5 LTD in the absence of postsynaptic spiking. J Neurophysiol, 2004. **92**(6): p. 3338–43.

Spanagel, R., A. Herz, and T.S. Shippenberg, Opposing tonically active endogenous opioid systems modulate the mesolimbic dopaminergic pathway. Proc Natl Acad Sci U S A, 1992. **89**(6): p. 2046–50.

Sugita, S., S.W. Johnson, and R.A. North, Synaptic inputs to GABAA and GABAB receptors originate from discrete afferent neurons. Neurosci Lett, 1992. **134**(2): p. 207–11.

Thomas, M.J., et al., Long-term depression in the nucleus accumbens: a neural correlate of behavioral sensitization to cocaine. Nat Neurosci, 2001. **4**(12): p. 1217–23.

Thomas, M.J. and R.C. Malenka, Synaptic plasticity in the mesolimbic dopamine system. Philos Trans R Soc Lond B Biol Sci, 2003. **358**(1432): p. 815–9.

Thomas, M.J., R.C. Malenka, and A. Bonci, Modulation of long-term depression by dopamine in the mesolimbic system. J Neurosci, 2000. **20**(15): p. 5581–6.

Ungless, M.A., P.J. Magill, and J.P. Bolam, Uniform inhibition of dopamine neurons in the ventral tegmental area by aversive stimuli. Science, 2004. **303**(5666): p. 2040–2.

Ungless, M.A., Dopamine: the salient issue. Trends Neurosci, 2004. **27**(12): p. 702–6.

Ungless, M.A., et al., Single cocaine exposure in vivo induces long-term potentiation in dopamine neurons. Nature, 2001. **411**(6837): p. 583–7.

Wang, T. and E.D. French, Electrophysiological evidence for the existence of NMDA and non-NMDA receptors on rat ventral tegmental dopamine neurons. Synapse, 1993. **13**(3): p. 270–7.

Wang, T. and E.D. French, L-glutamate excitation of A10 dopamine neurons is preferentially mediated by activation of NMDA receptors: extra- and intracellular electrophysiological studies in brain slices. Brain Res, 1993. **627**(2): p. 299–306.

Weiss, F., et al., Ethanol self-administration restores withdrawal-associated deficiencies in accumbal dopamine and 5-hydroxytryptamine release in dependent rats. J Neurosci, 1996. **16**(10): p. 3474–85.

Williams, J.T., M.J. Christie, and O. Manzoni, Cellular and synaptic adaptations mediating opioid dependence. Physiol Rev, 2001. **81**(1): p. 299–343.

Wise, R.A., Brain reward circuitry: insights from unsensed incentives. Neuron, 2002. **36**(2): p. 229–40.

Wise, R.A., Roles for nigrostriatal-not just mesocorticolimbic-dopamine in reward and addiction. Trends Neurosci, 2009.

Wolf, M.E., Addiction: making the connection between behavioral changes and neuronal plasticity in specific pathways. Mol Interv, 2002. **2**(3): p. 146–57.
Wolf, M.E., LTP may trigger addiction. Mol Interv, 2003. **3**(5): p. 248–52.
Wolf, M.E., The role of excitatory amino acids in behavioral sensitization to psychomotor stimulants. Prog Neurobiol, 1998. **54**(6): p. 679–720.
Xi, Z.X. and E.A. Stein, Nucleus accumbens dopamine release modulation by mesolimbic GABAA receptors-an in vivo electrochemical study. Brain Res, 1998. **798**(1–2): p. 156–65.
Xi, Z.X. and E.A. Stein, GABAergic mechanisms of opiate reinforcement. Alcohol Alcohol, 2002. **37**(5): p. 485–94.
Xi, Z.X. and E.A. Stein, Increased mesolimbic GABA concentration blocks heroin self-administration in the rat. J Pharmacol Exp Ther, 2000. **294**(2): p. 613–9.
Xi, Z.X. and E.A. Stein, Baclofen inhibits heroin self-administration behavior and mesolimbic dopamine release. J Pharmacol Exp Ther, 1999. **290**(3): p. 1369–74.
Yasuda, H., Y. Huang, and T. Tsumoto, Regulation of excitability and plasticity by endocannabinoids and PKA in developing hippocampus. Proc Natl Acad Sci U S A, 2008. **105**(8): p. 3106–11.
Zweifel, L.S., et al., Role of NMDA receptors in dopamine neurons for plasticity and addictive behaviors. Neuron, 2008. **59**(3): p. 486–96.
Zhuo, M., et al., Role of guanylyl cyclase and cGMP-dependent protein kinase in long-term potentiation. Nature, 1994. **368**(6472): p. 635–9.

Chapter 6
Ethanol Modulation of GABAergic Inhibition in Midbrain Dopamine Neurons: Implications for the Development of Alcohol-Seeking Behaviors

Jonathan W. Theile, Rueben A. Gonzales, and Richard A. Morrisett

Abstract Activation of dopaminergic (DA) neurons of the ventral tegmental area (VTA) by ethanol has been implicated in its rewarding and reinforcing effects. However, studies from our lab demonstrate that acute ethanol enhances GABA release onto VTA-DA neurons via activation of a G protein-coupled receptor, 5-HT2C, and subsequent activation of an intracellular calcium signaling pathway. Utilizing electrophysiological methods, in this chapter we present evidence which attempts to resolve the paradoxical nature of the dual excitatory and inhibitory actions of ethanol on DA neurons. Our results suggest that ethanol-excitation of VTA-DA neurons is bi-phasic and involves interplay of excitatory and inhibitory mechanisms which can fine tune the overall action of ethanol on DA neuron excitability. Overall, these results may provide insight into mechanisms underlying the development of alcohol dependence.

Preface

Addiction or drug dependence is generally defined as persistent and uncontrolled abuse of a substance despite serious negative consequences. Development of dependence and abuse of alcohol, one of the most widely abused drugs in society, is generally believed to be due to two major processes: (1) a genetic predisposition whereby 40–60% of the vulnerability to dependence can be attributed to genetic factors (Goldman et al. 2005; Hiroi and Agatsuma 2005) and (2) an abuse component whereby excessive drug use results in neuroadaptative changes which are thought to underlie the progression to and expression of the dependent or addicted state. Ethanol use dates back centuries and while most of us have become well aware of the behavioral, rewarding and cognitive effects of alcohol, we still do not

R.A. Morrisett (✉)
Institute for Cellular and Molecular Biology (Neurobiology) and the College of Pharmacy, University of Texas at Austin, 1 Univ Station, A1915, Austin, TX 78712, USA
e-mail: ramorris@mail.utexas.edu

M.A. Woodin and A. Maffei (eds.), *Inhibitory Synaptic Plasticity*,
DOI 10.1007/978-1-4419-6978-1_6,

fully understand the mechanisms by which chronic use and abuse of ethanol produces the critical neuroadaptive changes important in alcoholism. The difficulty in elucidating the mechanisms by which ethanol produces its effects are most likely related to the fact that ethanol has widespread effects throughout the brain which are not mediated by any single, specific ethanol receptor. Rather ethanol acts at numerous cellular proteins including, but not limited to, membrane-localized voltage- and ligand-gated ion channels as well as intracellular proteins and enzymes which modulate a myriad of signal transduction and cell signaling pathways.

Alcohol, the Mesocorticolimbic System and Neural Substrates of Addiction

One of the most critical neural pathways involved in processing an organism's response to rewarding stimuli is formed by the dopaminergic (DA) neurons of the ventral tegmental area (VTA) of the mesocorticolimbic dopamine system which send projections to the nucleus accumbens (NAc), prefrontal cortex (PFC), basolateral amygdala and a variety of other corticolimbic structures also involved in reward processing (Albanese and Minciacchi 1983; Oades and Halliday 1987). Most drugs of abuse, including ethanol, activate VTA-DA neurons via a variety of mechanisms culminating in aberrant release of DA onto these targets. Such pathological activation of the mesolimbic pathway is considered a primary step in the development and expression of alcoholism. The VTA consists primarily of DA and non-DA (presumed GABAergic) cell bodies which project to and receive inputs from these mesocorticolimbic structures. The VTA also receives reciprocal excitatory and inhibitory innervations from multiple brain regions. Glutamatergic inputs to the VTA arise from the medial PFC (mPFC), subthalamic nucleus, and the pedunculopontine nucleus (Christie et al. 1985; Groenewegen and Berendse 1990; Sesack and Pickel 1992; Charara et al. 1996). GABAergic neurons from the NAc and ventral pallidum project to VTA (Walaas and Fonnum 1980), and VTA-GABAergic neurons project back to both the mPFC (Carr and Sesack 2000) and the NAc (Van Bockstaele and Pickel 1995). Additionally, local VTA GABAergic neurons (interneurons) project onto neighboring VTA-DA cells and are thought to regulate the activity of these DA cells via tonic inhibition (Yim and Mogenson 1980; Johnson and North 1992a, b; Westerink et al. 1996).

It is well established that dopamine transmission within the mesocorticolimbic system is a critical component in the regulation of alcohol drinking behaviors. Systemic administration of dopamine agonists and antagonists reduced alcohol drinking in rats (Pfeffer and Samson 1986; Pfeffer and Samson 1988) and local microinjection of dopamine antagonists and agonists into the NAc terminated and prolonged ethanol drinking, respectively (Samson et al. 1992). Furthermore, in vivo microdialysis studies have demonstrated an increase in dopamine levels in the NAc in P and Wistar rats prior to and during ethanol self-administration (Weiss et al. 1993; Weiss et al. 1996; Gonzales and Weiss 1998; Melendez et al. 2002).

Acute ethanol administration at intoxicating doses has been shown to increase the spontaneous firing rate of DA neurons in the VTA both in vivo (Gessa et al. 1985) and in vitro (Brodie et al. 1990). The latter study demonstrated a dose-dependent increase in spontaneous firing rate that was assumed to be largely attributable to a direct postsynaptic effect since the recordings were conducted under conditions that block synaptic transmission. Additionally, the firing rate of acutely dissociated VTA-DA neurons was also enhanced by ethanol which demonstrates that ethanol may enhance DA neuron excitability even in the absence of synaptic input (Brodie et al. 1999). Recently, several ethanol targets have emerged giving insight into the potential mechanism(s) by which ethanol stimulates VTA-DA neuron activity. In C57BL/6J mice, ethanol enhanced the hyperpolarization-activated inward current, I_h, and thus may contribute to the excitatory effects on VTA-DA firing via actions at that channel (Okamoto et al. 2006). Additionally, ethanol enhanced firing rate through inhibition of a quinidine-sensitive delayed rectifier K^+ current (Appel et al. 2003) and the voltage-gated K^+ current, I_M (Koyama et al. 2007). These results suggest that ethanol enhances spontaneous firing frequency through modulation of ion channels that regulate pacemaking activity.

Ethanol and Opioids

The rewarding and reinforcing properties of ethanol have also been linked to various opioid receptor subtypes in the mesolimbic system. Opioid receptors are G protein-coupled receptors (GPCRs) that utilize multiple transduction mechanisms. Three known classes of opioid receptors include mu, delta and kappa which show high affinity for the endogenous opioids beta-endorphin, enkephalin and dynorphin, respectively (Zollner and Stein 2007). Acute ethanol exposure has been demonstrated to increase endogenous beta-endorphin levels in a variety of brain structures (Gianoulakis and Barcomb 1987; de Waele and Gianoulakis 1993; Olive et al. 2001) including the midbrain (Rasmussen et al. 1998; Jarjour et al. 2009). Ethanol also increased levels of enkephalins and dynorphins but results between studies have been inconsistent (Jarjour et al. 2009). Naltrexone and naloxone, two antagonists with affinity for all three opioid receptor subtypes, reduce ethanol consummatory behaviors (Gonzales and Weiss 1998; Shoemaker et al. 2002; Stromberg et al. 2002). However, there is more consistent evidence demonstrating mu-opioid receptor (MOR) modulation of ethanol administration. The use of selective MOR antagonists or genetic deletion of MORs reduces ethanol drinking (Honkanen et al. 1996; Krishnan-Sarin et al. 1998; Stromberg et al. 1998; Kim et al. 2000; Hall et al. 2001; Hyytia and Kiianmaa 2001; Mhatre and Holloway 2003; Lasek et al. 2007; Margolis et al. 2008).

Ethanol intake increases the output of VTA-DA neurons, and this output is concurrently modulated by opioidergic systems. Recent evidence suggests that the ethanol-induced increase in dopamine release to the NAc may be modulated by MORs (Job et al. 2007). In that study, pharmacological blockade with the MOR-selective antagonist, naloxonazine, or genetic deletion of MORs reduced the ethanol-mediated enhancement of DA release under some, but not all conditions, thereby suggesting that MORs may regulate DA activity in the VTA. Indeed, MORs are localized presynaptically to GABAergic terminals in the VTA (Svingos et al. 2001) and activation of these receptors disinhibits DA neurons (Johnson and North 1992a; Margolis et al. 2003). In light of this evidence, ethanol may decrease GABAergic transmission via MOR-mediated inhibition resulting in disinhibition of DA neurons thereby constituting a potential mechanism for MOR regulation of ethanol action.

Ethanol and GABAergic Transmission

There is a general agreement that acutely administered ethanol facilitates presynaptic GABA release in numerous brain regions including the cerebellum (Ming et al. 2006; Kelm et al. 2007), hippocampus (Ariwodola and Weiner 2004; Sanna et al. 2004) and amygdala (Roberto et al. 2003; Zhu and Lovinger 2006). Surprisingly, the notion that ethanol could modulate GABA release went largely unexplored after early reports in the 1980s demonstrated an ethanol-enhancement of postsynaptic $GABA_A$ receptor function. Nearly 20 years would pass before the emergence of clear evidence of a potentiating effect of ethanol on presynaptic GABA release. Several electrophysiological methods can be used to detect changes in inhibitory neurotransmitter release. Two methods commonly used for in vitro slice preparations include the analysis of quantal, miniature (action potential independent; tetrodotoxin-insensitive) and spontaneous (action potential-dependent; tetrodotoxin-sensitive) inhibitory postsynaptic currents (m/sIPSCs). A change in the frequency of m/sIPSCs reflects a change in the number of release events; a change in the amplitude of m/sIPSCs represents a change in postsynaptic receptor sensitivity. Another method to study GABA release is by analyzing the change in the ratio of paired evoked IPSCs (paired-pulse ratio).

Given the important role of the midbrain in the development and expression of alcohol dependence, there are surprisingly few studies exploring ethanol modulation of GABA release in the VTA. The first electrophysiological study to measure the effect of ethanol on GABA release in this brain region demonstrated that 24 h after a single injection of ethanol in mice, there was a long-lasting potentiation of GABA release that was interpreted as a manifestation of a long-lasting effect of acute ethanol (Melis et al. 2002; Wanat et al. 2008). Conversely, Ye and colleagues demonstrated that ethanol (40 mM) substantially inhibited the firing rate of presumed VTA-GABAergic interneurons in vitro (Xiao et al. 2007). In that study, these inhibitory interneurons, and their VTA-DA targets, were under strong opioid influence as demonstrated by a marked increased in DA neuron firing rate by the MOR agonist, DAMGO, and a decrease in firing rate by the opioid antagonist, naloxone. Additionally, ethanol decreased action potential-dependent mediated GABA transmission onto VTA-DA neurons (Xiao and Ye 2008). Interestingly, in the presence of DAMGO an enhancement in GABA transmission was observed, which was attributed to ethanol activation of GABAergic projections from other brain regions following silencing of local interneurons via MOR activation. Nevertheless, under no conditions did Ye and colleagues observe any ethanol-induced change in mIPSC frequency (Xiao and Ye 2008). Steffensen and colleagues have demonstrated ethanol inhibition of interneuron excitability believed to be mediated through inhibition of NMDA receptors (Stobbs et al. 2004). Interestingly, they also demonstrated that low dose intravenous ethanol (0.01–0.03 g/kg) enhanced VTA-GABA neuron firing rate in rats (Steffensen et al. 2009). The authors contend in the latter study that ethanol enhanced dopamine release which then acted to excite local GABA interneurons. Thus conflicting results regarding the effects of ethanol on GABA release in the VTA demands the need for further investigation.

In light of those disparate observations and lack of understanding of the efficacy of mu-opiate receptor antagonists in ethanol dependence, we sought to investigate ethanol modulation of GABAergic transmission in the VTA. In our first report (Theile et al. 2008) we demonstrated that ethanol (25 and 50 mM) significantly enhanced sIPSC frequency and amplitude. Furthermore, ethanol (50 mM) enhanced paired-pulse depression and mIPSC frequency, indicative of an enhancement in GABA release onto VTA-DA neurons. While other investigators, working in the hippocampus, observed that $GABA_B$ receptor activity modulates ethanol-enhancement of $GABA_A$ receptor-mediated IPSCs (Ariwodola and Weiner 2004), the same was not the case in the VTA. We observed a decrease in sIPSC frequency mediated by $GABA_B$ receptor activation which confirmed previous findings (Rohrbacher et al. 1997). Consistent with the concept of $GABA_B$ receptor-mediated auto-inhibition of GABAergic neurons, we observed that $GABA_B$ receptor blockade caused an increase in sIPSC frequency. Blockade of $GABA_B$ receptors presumably removed this tonic auto-inhibition, allowing for increased GABA release. Nevertheless, neither activation nor blockade of $GABA_B$ receptors appeared to affect ethanol enhancement of sIPSC frequency, indeed indicating that ethanol modulation of GABA release is independent of $GABA_B$ receptor function. In a neighboring region of the VTA, the substantia nigra pars reticulata (SNr), ethanol (100 mM) has also been observed to enhance $GABA_A$ mIPSC frequency (Criswell et al. 2008).

In order to study ethanol-VTA GABA interactions more definitively, we initiated studies of mIPSCs. In the second report from our lab (Theile et al. 2009) we further characterized the mechanism of ethanol-enhancement in GABA release. Another group had recently observed ethanol caused an increase in GABA release onto cerebellar Purkinje neurons by activation of calcium release from presynaptic inositol-1-4-5-triphosphate (IP_3)- and ryanodine–sensitive internal stores (Kelm et al. 2007). Likewise, we observed a similar requirement for ethanol potentiation of mIPSCs as recorded from VTA-DA neurons. Blockade of IP_3 receptors and depletion of intracellular stores both occluded the ethanol-enhancement in mIPSC frequency. Coincidently we became aware that a GPCR coupled to IP_3 production stimulated by serotonin ($5\text{-}HT_{2C}$) had been demonstrated to exist on VTA GABA neurons and modulate VTA-DA activity (Di Giovanni et al. 1999; Di Matteo et al. 2000; Bubar and Cunningham 2007). Consequently, we initiated studies of $5\text{-}HT_{2C}$

receptor modulation of GABA transmission in the VTA. We demonstrated a robust and long-lasting 5-HT_{2C} receptor-dependent enhancement in GABA release onto VTA-DA neurons. Furthermore, blockade of 5-HT_{2C} receptors occluded the agonist-induced as well as the ethanol-enhancement in GABA release. Therefore, our data directly demonstrate that 5-HT_{2C}/IP_3 receptor activation and subsequent calcium mobilization represents a primary candidate for the mechanism by which acute ethanol enhances GABA release in the VTA. Ethanol increases 5-HT release in the NAc (Yoshimoto et al. 1992) and prevents 5-HT reuptake in the hippocampus (Daws et al. 2006). Therefore, it is conceivable that ethanol may also stimulate 5-HT release in the VTA which subsequently results in the activation of 5-HT_{2C}Rs. This is certainly a possibility given that application of 5-HT enhances GABA release in our preparation as well (Theile et al. 2009).

Our most recent work has further explored the mechanisms of ethanol-enhancement of VTA-DA neuron activity in light of our recent findings demonstrating a concurrent enhancement in GABA release in the VTA. Since ethanol might have opposing actions in the VTA to enhance inhibitory drive from VTA-GABA neurons while simultaneously enhancing excitability of VTA-DA neurons, one might expect complex actions of ethanol on VTA-DA firing patterns. Indeed, ethanol (75 mM) application did elicit a robust and reliable increase in VTA-DA neuron firing rate that consisted of two distinct phases: a rising and falling phase (Fig. 6.1a). Immediately after ethanol was introduced into the bath, we consistently observed a rapid increase in spontaneous firing rate that peaked and then decreased somewhat before stabilizing at a level intermediate between the baseline and peak responses. We interpreted these results as indicating that the excitatory effect of ethanol on VTA-DA excitation is not maintained throughout the entire duration of ethanol exposure, but rather is affected by another slower process which acts to diminish the excitation to the degree of decreasing the firing rate to baseline levels. We hypothesized that an ethanol-induced enhancement in inhibitory drive may constitute a primary candidate mechanism that dampens the excitatory effect of ethanol on VTA-DA firing rate. Therefore, we tested whether the time-course of ethanol enhancement of VTA-DA firing rate was altered in the presence of saturating concentrations of picrotoxin, a GABA channel blocker. Indeed, we observed distinct differences in the kinetics of ethanol actions on VTA-DA firing under conditions of GABA block such that the peak excitatory effect of ethanol was larger and sustained for a longer period of time (Fig. 6.1). Furthermore, while the $GABA_A$ receptor agonist muscimol strongly inhibited VTA-DA firing rate, this inhibition was more pronounced in the presence of ethanol (Fig. 6.2). Thus, while ethanol-enhancement of GABA release alone is not enough to overcome the direct excitatory effect of ethanol on VTA-DA neuron activity, these results demonstrate that conditions of adjunctive $GABA_A$ receptor activation can reverse the stimulatory effect of ethanol on VTA-DA neuron activity.

Several labs have demonstrated a MOR-mediated disinhibition of VTA-DA neurons (Johnson and North 1992a; Margolis et al. 2003; Xiao et al. 2007). It should be noted, however, that in the study by Margolis et al. (2003), disinhibition was seen in only 47% of the neurons tested. Additionally, in the study by Johnson and North

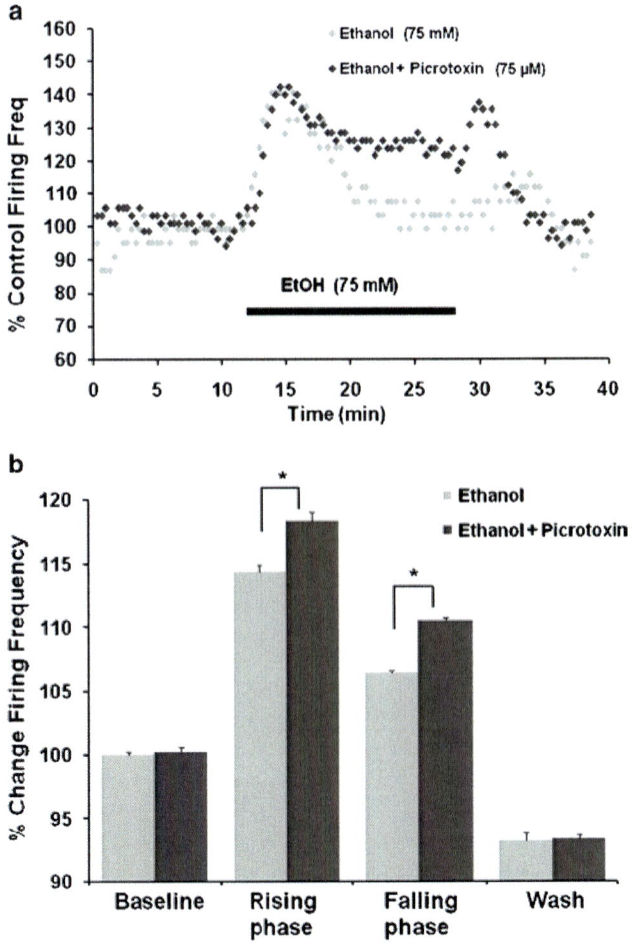

Fig. 6.1 $GABA_A$ receptor blockade sustains ethanol-enhancement in VTA-DA firing activity. (**a**) Sample neurons showing the normalized data points in the presence of ethanol (75 mM) alone and in the continued presence of picrotoxin (75 μM). Data points for individual cells represent 20-s sweeps in which the average firing rate (Hz) was calculated. Each data point is normalized to the last 5 min of baseline. (**b**) A bar graph representing the % change ± SEM above control firing rate. The bar graph represents the average 5 min ethanol response (peak rising phase and peak falling phase) and the last 5 min of washout normalized to the last 5 min of the baseline (control). (ethanol; $n=30$, ethanol + picrotoxin; $n=14$, * indicates $p<0.01$ by one-way ANOVA)

(1992a), the use of an elevated extracellular K^+ solution was required to observe disinhibition. However, we did not observe a MOR-mediated enhancement in either spontaneous or ethanol-enhanced firing rate. Although DAMGO reduced basal GABA release, it did not occlude the ethanol-enhancement in release; thus, the mechanism of ethanol modulation of GABA release is clearly distinct from that of opioid modulation. Naltrexone has been demonstrated to reduce the ethanol-induced

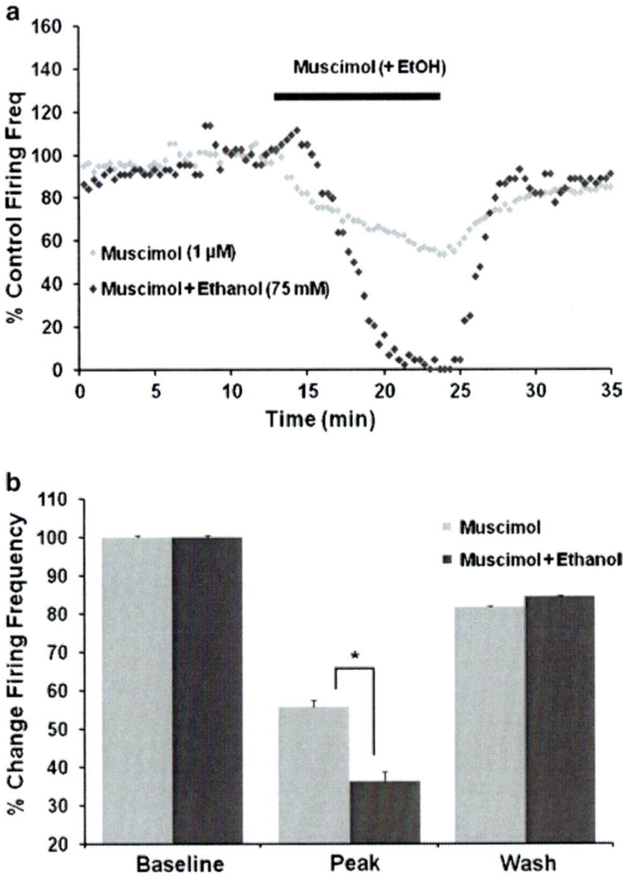

Fig. 6.2 $GABA_A$ receptor activation uncovers inhibitory effect of ethanol on VTA-DA firing activity. (**a**) Sample neurons showing the normalized data points in the presence of muscimol (1 μM) alone and with co-application of muscimol and ethanol (75 mM). Data points for individual cells represent 20-s sweeps in which the average firing rate (Hz) was calculated. Each data point is normalized to the last 5 min of baseline. (**b**) A bar graph representing the % change ± SEM below control firing rate. The bar graph represents the peak 5 min muscimol effect on inhibition (and muscimol with ethanol) and the last 5 min of washout normalized to the last 5 min of the baseline (control). (muscimol; $n=8$, muscimol + ethanol; $n=9$, * indicates $p<0.01$ by one-way ANOVA)

enhancement in dopamine release from the VTA (Gonzales and Weiss 1998). However, we did not observe a reduction in the ethanol-enhancement in firing rate in the presence of naltrexone. Thus, we do not observe in vitro the ability of naltrexone to suppress the ethanol enhancement in VTA-DA neuron activity as seen in vivo. Opioid tone is likely substantially reduced or absent in our slice preparation due to deafferentation, one of the few notable drawbacks to studying synaptic transmission in an in vitro slice preparation.

Implications for Ethanol Modulation of Dopamine Output

An important question that remains concerns the relevance of our recent findings to the mechanism of alcohol dependence. This question can be better answered by studying the effects of both acute and chronic ethanol on synaptic transmission in the VTA. Therefore, any forthcoming conclusions must reflect the fact that we have no information concerning the effects of chronic ethanol exposure on GABA release in the VTA. However, one study did demonstrate that VTA-DA neurons are sensitized to ethanol after chronic exposure and the sensitivity of DA neurons to GABA was reduced (Brodie 2002). Whether or not decreased sensitivity to GABA reflects any compensatory changes in GABAergic transmission remains uninvestigated, although variable changes in $GABA_A$ receptor expression in different brain regions following exposure to chronic ethanol have been documented (Enoch 2008; Kumar et al. 2009).

Obviously, ethanol is used intermittently and our understanding of changes in VTA-DA and GABA interactions during withdrawal is unknown. Several studies have examined the effects of long-term ethanol administration on GABAergic transmission; however, there is a lack of consensus between brain regions and in the type of ethanol administration paradigm used. In the hippocampus, withdrawal from chronic intermittent ethanol (CIE) treatment results in a decrease in mIPSC frequency as recorded from CA1 neurons (Cagetti et al. 2003). Although subsequent exposure to acute ethanol was not examined in that study, a previous study in the hippocampus demonstrated an enhancement in ethanol sensitivity to evoked IPSCs (Kang et al. 1998). Conversely, in the amygdala, continuous chronic ethanol exposure results in an enhancement in GABA release (Roberto et al. 2004). Furthermore, in that study acute ethanol similarly enhanced GABA release in both chronic treated and ethanol-naïve rats, thus suggesting a lack of tolerance to acute ethanol. Nonetheless, based on the results presented by Brodie (2002) there must be some factor regulating the altered sensitivity of VTA-DA neurons to $GABA_A$ receptor-mediated inhibition. Prolonged GABA release induced by ethanol may impart compensatory changes in $GABA_A$ receptor expression and/or function to account for the results seen by Brodie. Indeed, chronic $GABA_A$ receptor activation can decrease $GABA_A$ receptor subunit expression and $GABA_A$ receptor-mediated inhibition (Roca et al. 1990; Montpied et al. 1991). Furthermore, Melis et al. (2002) demonstrated that a single in vivo injection of ethanol resulted in a long-lasting enhancement in GABA release. Thus it is possible that ethanol-induced changes in GABA release could contribute to the altered VTA-DA sensitivity to GABA-mediated inhibition. As a result, it would be pertinent to investigate the effects of long-term ethanol exposure on GABA release in the VTA, in vitro and in vivo, using multiple ethanol administration paradigms.

The $GABA_A$ receptor dependent switch from ethanol stimulation to inhibition may have implications for the rewarding and reinforcing properties of ethanol. In the whole animal, strong ethanol stimulation of brain regions with GABAergic projections to the VTA in conjunction with stimulation of local

GABAergic interneurons may strongly impact the direct excitatory effect of ethanol on VTA-DA neuron activity. Initial sensitivity to ethanol is a strong biological factor predicative to the development of dependence (Schuckit et al. 1985). Thus, in those that become dependent, basal and/or ethanol-enhanced GABA release may initially be minimal or become diminished after chronic exposure. This would correlate to a stronger excitatory effect of ethanol on VTA-DA neuron activity and thus ethanol consumption could become aberrantly reinforcing. Conversely, those who are less susceptible to develop alcohol dependence may exhibit robust basal and/or ethanol-enhanced GABA release, thereby curbing the excitatory effect of ethanol on VTA-DA neuron activity. This in turn may limit the reinforcing actions of ethanol within the mesolimbic system. It would be very interesting to investigate potential differences in the acute effects of ethanol on GABA release between strains of animals that are genetically pre-disposed to drink more or less. We predict that animal strains which are genetically prone to drink less or not at all (alcohol non-preferring) may exhibit a strong basal GABA tone and/or a robust ethanol-enhancement in GABA release which could correlate with a weaker reinforcing effect of acute ethanol. Conversely, those strains which are genetically prone to drink more (alcohol preferring) may exhibit a weak GABA tone and/or a weaker ethanol-enhancement in GABA release and thus exhibit a strong reinforcing effect of ethanol.

Acknowledgements This work was supported by the following grants: R01 AA 14874 (RAG), U01 AA 16651 (RAM), F31 AA 17020 (JWT).

References

Albanese A and Minciacchi D (1983) Organization of the ascending projections from the ventral tegmental area: a multiple fluorescent retrograde tracer study in the rat. *J Comp Neurol* **216**:406–420.

Appel SB, Liu Z, McElvain MA and Brodie MS (2003) Ethanol excitation of dopaminergic ventral tegmental area neurons is blocked by quinidine. *J Pharmacol Exp Ther* **306**:437–446.

Ariwodola OJ and Weiner JL (2004) Ethanol potentiation of GABAergic synaptic transmission may be self-limiting: role of presynaptic GABA(B) receptors. *J Neurosci* **24**:10679–10686.

Brodie MS (2002) Increased ethanol excitation of dopaminergic neurons of the ventral tegmental area after chronic ethanol treatment. *Alcohol Clin Exp Res* **26**:1024–1030.

Brodie MS, Pesold C and Appel SB (1999) Ethanol directly excites dopaminergic ventral tegmental area reward neurons. *Alcohol Clin Exp Res* **23**:1848–1852.

Brodie MS, Shefner SA and Dunwiddie TV (1990) Ethanol increases the firing rate of dopamine neurons of the rat ventral tegmental area in vitro. *Brain Res* **508**:65–69.

Bubar MJ and Cunningham KA (2007) Distribution of serotonin 5-HT2C receptors in the ventral tegmental area. *Neuroscience* **146**:286–297.

Cagetti E, Liang J, Spigelman I and Olsen RW (2003) Withdrawal from chronic intermittent ethanol treatment changes subunit composition, reduces synaptic function, and decreases behavioral responses to positive allosteric modulators of GABAA receptors. *Mol Pharmacol* **63**:53–64.

Carr DB and Sesack SR (2000) GABA-containing neurons in the rat ventral tegmental area project to the prefrontal cortex. *Synapse* **38**:114–123.

Charara A, Smith Y and Parent A (1996) Glutamatergic inputs from the pedunculopontine nucleus to midbrain dopaminergic neurons in primates: Phaseolus vulgaris-leucoagglutinin anterograde labeling combined with postembedding glutamate and GABA immunohistochemistry. *J Comp Neurol* **364**:254–266.

Christie MJ, Bridge S, James LB and Beart PM (1985) Excitotoxin lesions suggest an aspartatergic projection from rat medial prefrontal cortex to ventral tegmental area. *Brain Res* **333**:169–172.

Criswell HE, Ming Z, Kelm MK and Breese GR (2008) Brain regional differences in the effect of ethanol on GABA release from presynaptic terminals. *J Pharmacol Exp Ther* **326**:596–603.

Daws LC, Montanez S, Munn JL, Owens WA, Baganz NL, Boyce-Rustay JM, Millstein RA, Wiedholz LM, Murphy DL and Holmes A (2006) Ethanol inhibits clearance of brain serotonin by a serotonin transporter-independent mechanism. *J Neurosci* **26**:6431–6438.

de Waele JP and Gianoulakis C (1993) Effects of single and repeated exposures to ethanol on hypothalamic beta-endorphin and CRH release by the C57BL/6 and DBA/2 strains of mice. *Neuroendocrinology* **57**:700–709.

Di Giovanni G, De Deurwaerdere P, Di Mascio M, Di Matteo V, Esposito E and Spampinato U (1999) Selective blockade of serotonin-2C/2B receptors enhances mesolimbic and mesostriatal dopaminergic function: a combined in vivo electrophysiological and microdialysis study. *Neuroscience* **91**:587–597.

Di Matteo V, Di Giovanni G, Di Mascio M and Esposito E (2000) Biochemical and electrophysiological evidence that RO 60-0175 inhibits mesolimbic dopaminergic function through serotonin(2C) receptors. *Brain Res* **865**:85–90.

Enoch MA (2008) The role of GABA(A) receptors in the development of alcoholism. *Pharmacol Biochem Behav* **90**:95–104.

Gessa GL, Muntoni F, Collu M, Vargiu L and Mereu G (1985) Low doses of ethanol activate dopaminergic neurons in the ventral tegmental area. *Brain Res* **348**:201–203.

Gianoulakis C and Barcomb A (1987) Effect of acute ethanol in vivo and in vitro on the beta-endorphin system in the rat. *Life Sci* **40**:19–28.

Goldman D, Oroszi G and Ducci F (2005) The genetics of addictions: uncovering the genes. *Nat Rev Genet* **6**:521–532.

Gonzales RA and Weiss F (1998) Suppression of ethanol-reinforced behavior by naltrexone is associated with attenuation of the ethanol-induced increase in dialysate dopamine levels in the nucleus accumbens. *J Neurosci* **18**:10663–10671.

Groenewegen HJ and Berendse HW (1990) Connections of the subthalamic nucleus with ventral striatopallidal parts of the basal ganglia in the rat. *J Comp Neurol* **294**:607–622.

Hall FS, Sora I and Uhl GR (2001) Ethanol consumption and reward are decreased in mu-opiate receptor knockout mice. *Psychopharmacology (Berl)* **154**:43–49.

Hiroi N and Agatsuma S (2005) Genetic susceptibility to substance dependence. *Mol Psychiatry* **10**:336–344.

Honkanen A, Vilamo L, Wegelius K, Sarviharju M, Hyytia P and Korpi ER (1996) Alcohol drinking is reduced by a mu 1- but not by a delta-opioid receptor antagonist in alcohol-preferring rats. *Eur J Pharmacol* **304**:7–13.

Hyytia P and Kiianmaa K (2001) Suppression of ethanol responding by centrally administered CTOP and naltrindole in AA and Wistar rats. *Alcohol Clin Exp Res* **25**:25–33.

Jarjour S, Bai L and Gianoulakis C (2009) Effect of Acute Ethanol Administration on the Release of Opioid Peptides From the Midbrain Including the Ventral Tegmental Area. *Alcohol Clin Exp Res* **33**:1033–1043.

Job MO, Tang A, Hall FS, Sora I, Uhl GR, Bergeson SE and Gonzales RA (2007) Mu (mu) opioid receptor regulation of ethanol-induced dopamine response in the ventral striatum: evidence of genotype specific sexual dimorphic epistasis. *Biol Psychiatry* **62**:627–634.

Johnson SW and North RA (1992a) Opioids excite dopamine neurons by hyperpolarization of local interneurons. *J Neurosci* **12**:483–488.

Johnson SW and North RA (1992b) Two types of neurone in the rat ventral tegmental area and their synaptic inputs. *J Physiol* **450**:455–468.

Kang MH, Spigelman I and Olsen RW (1998) Alteration in the sensitivity of GABA(A) receptors to allosteric modulatory drugs in rat hippocampus after chronic intermittent ethanol treatment. *Alcohol Clin Exp Res* **22**:2165–2173.

Kelm MK, Criswell HE and Breese GR (2007) Calcium release from presynaptic internal stores is required for ethanol to increase spontaneous gamma-aminobutyric acid release onto cerebellum Purkinje neurons. *J Pharmacol Exp Ther* **323**:356–364.

Kim SG, Stromberg MF, Kim MJ, Volpicelli JR and Park JM (2000) The effect of antagonists selective for mu- and delta-opioid receptor subtypes on alcohol consumption in C57BL/6 mice. *Alcohol* **22**:85–90.

Koyama S, Brodie MS and Appel SB (2007) Ethanol inhibition of m-current and ethanol-induced direct excitation of ventral tegmental area dopamine neurons. *J Neurophysiol* **97**:1977–1985.

Krishnan-Sarin S, Wand GS, Li XW, Portoghese PS and Froehlich JC (1998) Effect of mu opioid receptor blockade on alcohol intake in rats bred for high alcohol drinking. *Pharmacol Biochem Behav* **59**:627–635.

Kumar S, Porcu P, Werner DF, Matthews DB, Diaz-Granados JL, Helfand RS and Morrow AL (2009) The role of GABA(A) receptors in the acute and chronic effects of ethanol: a decade of progress. *Psychopharmacology (Berl)* **205**:529–564.

Lasek AW, Janak PH, He L, Whistler JL and Heberlein U (2007) Downregulation of mu opioid receptor by RNA interference in the ventral tegmental area reduces ethanol consumption in mice. *Genes Brain Behav* **6**:728–735.

Margolis EB, Fields HL, Hjelmstad GO and Mitchell JM (2008) Delta-opioid receptor expression in the ventral tegmental area protects against elevated alcohol consumption. *J Neurosci* **28**:12672–12681.

Margolis EB, Hjelmstad GO, Bonci A and Fields HL (2003) Kappa-opioid agonists directly inhibit midbrain dopaminergic neurons. *J Neurosci* **23**:9981–9986.

Melendez RI, Rodd-Henricks ZA, Engleman EA, Li TK, McBride WJ and Murphy JM (2002) Microdialysis of dopamine in the nucleus accumbens of alcohol-preferring (P) rats during anticipation and operant self-administration of ethanol. *Alcohol Clin Exp Res* **26**:318–325.

Melis M, Camarini R, Ungless MA and Bonci A (2002) Long-lasting potentiation of GABAergic synapses in dopamine neurons after a single in vivo ethanol exposure. *J Neurosci* **22**:2074–2082.

Mhatre M and Holloway F (2003) Micro1-opioid antagonist naloxonazine alters ethanol discrimination and consumption. *Alcohol* **29**:109–116.

Ming Z, Criswell HE, Yu G and Breese GR (2006) Competing presynaptic and postsynaptic effects of ethanol on cerebellar purkinje neurons. *Alcohol Clin Exp Res* **30**:1400–1407.

Montpied P, Ginns EI, Martin BM, Roca D, Farb DH and Paul SM (1991) gamma-Aminobutyric acid (GABA) induces a receptor-mediated reduction in GABAA receptor alpha subunit messenger RNAs in embryonic chick neurons in culture. *J Biol Chem* **266**:6011–6014.

Oades RD and Halliday GM (1987) Ventral tegmental (A10) system: neurobiology. 1. Anatomy and connectivity. *Brain Res* **434**:117–165.

Okamoto T, Harnett MT and Morikawa H (2006) Hyperpolarization-activated cation current (Ih) is an ethanol target in midbrain dopamine neurons of mice. *J Neurophysiol* **95**:619–626.

Olive MF, Koenig HN, Nannini MA and Hodge CW (2001) Stimulation of endorphin neurotransmission in the nucleus accumbens by ethanol, cocaine, and amphetamine. *J Neurosci* **21**:RC184.

Pfeffer AO and Samson HH (1986) Effect of pimozide on home cage ethanol drinking in the rat: dependence on drinking session length. *Drug Alcohol Depend* **17**:47–55.

Pfeffer AO and Samson HH (1988) Haloperidol and apomorphine effects on ethanol reinforcement in free feeding rats. *Pharmacol Biochem Behav* **29**:343–350.

Rasmussen DD, Bryant CA, Boldt BM, Colasurdo EA, Levin N and Wilkinson CW (1998) Acute alcohol effects on opiomelanocortinergic regulation. *Alcohol Clin Exp Res* **22**:789–801.

Roberto M, Madamba SG, Moore SD, Tallent MK and Siggins GR (2003) Ethanol increases GABAergic transmission at both pre- and postsynaptic sites in rat central amygdala neurons. *Proc Natl Acad Sci U S A* **100**:2053–2058.

Roberto M, Madamba SG, Stouffer DG, Parsons LH and Siggins GR (2004) Increased GABA release in the central amygdala of ethanol-dependent rats. *J Neurosci* **24**:10159–10166.

Roca DJ, Schiller GD, Friedman L, Rozenberg I, Gibbs TT and Farb DH (1990) gamma-Aminobutyric acidA receptor regulation in culture: altered allosteric interactions following prolonged exposure to benzodiazepines, barbiturates, and methylxanthines. *Mol Pharmacol* **37**:710–719.

Rohrbacher J, Jarolimek W, Lewen A and Misgeld U (1997) GABAB receptor-mediated inhibition of spontaneous inhibitory synaptic currents in rat midbrain culture. *J Physiol* **500 (Pt 3)**:739–749.

Samson HH, Tolliver GA, Haraguchi M and Hodge CW (1992) Alcohol self-administration: role of mesolimbic dopamine. *Ann N Y Acad Sci* **654**:242–253.

Sanna E, Talani G, Busonero F, Pisu MG, Purdy RH, Serra M and Biggio G (2004) Brain steroidogenesis mediates ethanol modulation of GABAA receptor activity in rat hippocampus. *J Neurosci* **24**:6521–6530.

Sesack SR and Pickel VM (1992) Prefrontal cortical efferents in the rat synapse on unlabeled neuronal targets of catecholamine terminals in the nucleus accumbens septi and on dopamine neurons in the ventral tegmental area. *J Comp Neurol* **320**:145–160.

Shoemaker WJ, Vavrousek-Jakuba E, Arons CD and Kwok FC (2002) The acquisition and maintenance of voluntary ethanol drinking in the rat: effects of dopaminergic lesions and naloxone. *Behav Brain Res* **137**:139–148.

Schuckit MA, Li TK, Cloninger CR, Deitrich RA (1985) Genetics of alcoholism. *Alcohol Clin Exp Res* **6**:475–92.

Steffensen SC, Walton CH, Hansen DM, Yorgason JT, Gallegos RA and Criado JR (2009) Contingent and non-contingent effects of low-dose ethanol on GABA neuron activity in the ventral tegmental area. *Pharmacol Biochem Behav* **92**:68–75.

Stobbs SH, Ohran AJ, Lassen MB, Allison DW, Brown JE and Steffensen SC (2004) Ethanol suppression of ventral tegmental area GABA neuron electrical transmission involves N-methyl-D-aspartate receptors. *J Pharmacol Exp Ther* **311**:282–289.

Stromberg MF, Casale M, Volpicelli L, Volpicelli JR and O'Brien CP (1998) A comparison of the effects of the opioid antagonists naltrexone, naltrindole, and beta-funaltrexamine on ethanol consumption in the rat. *Alcohol* **15**:281–289.

Stromberg MF, Sengpiel T, Mackler SA, Volpicelli JR, O'Brien CP and Vogel WH (2002) Effect of naltrexone on oral consumption of concurrently available ethanol and cocaine in the rat. *Alcohol* **28**:169–179.

Svingos AL, Garzon M, Colago EE and Pickel VM (2001) Mu-opioid receptors in the ventral tegmental area are targeted to presynaptically and directly modulate mesocortical projection neurons. *Synapse* **41**:221–229.

Theile JW, Morikawa H, Gonzales RA and Morrisett RA (2008) Ethanol enhances GABAergic transmission onto dopamine neurons in the ventral tegmental area of the rat. *Alcohol Clin Exp Res* **32**:1040–1048.

Theile JW, Morikawa H, Gonzales RA and Morrisett RA (2009) Role of 5-hydroxytryptamine2C receptors in Ca2+-dependent ethanol potentiation of GABA release onto ventral tegmental area dopamine neurons. *J Pharmacol Exp Ther* **329**:625–633.

Van Bockstaele EJ and Pickel VM (1995) GABA-containing neurons in the ventral tegmental area project to the nucleus accumbens in rat brain. *Brain Res* **682**:215–221.

Walaas I and Fonnum F (1980) Biochemical evidence for gamma-aminobutyrate containing fibres from the nucleus accumbens to the substantia nigra and ventral tegmental area in the rat. *Neuroscience* **5**:63–72.

Wanat MJ, Sparta DR, Hopf FW, Bowers MS, Melis M and Bonci A (2008) Strain Specific Synaptic Modifications on Ventral Tegmental Area Dopamine Neurons After Ethanol Exposure. *Biol Psychiatry* **65(8)**:646–53.

Weiss F, Lorang MT, Bloom FE and Koob GF (1993) Oral alcohol self-administration stimulates dopamine release in the rat nucleus accumbens: genetic and motivational determinants. *J Pharmacol Exp Ther* **267**:250–258.

Weiss F, Parsons LH, Schulteis G, Hyytia P, Lorang MT, Bloom FE and Koob GF (1996) Ethanol self-administration restores withdrawal-associated deficiencies in accumbal dopamine and 5-hydroxytryptamine release in dependent rats. *J Neurosci* **16**:3474–3485.

Westerink BH, Kwint HF and deVries JB (1996) The pharmacology of mesolimbic dopamine neurons: a dual-probe microdialysis study in the ventral tegmental area and nucleus accumbens of the rat brain. *J Neurosci* **16**:2605–2611.

Xiao C and Ye JH (2008) Ethanol dually modulates GABAergic synaptic transmission onto dopaminergic neurons in ventral tegmental area: role of mu-opioid receptors. *Neuroscience* **153**:240–248.

Xiao C, Zhang J, Krnjevic K and Ye JH (2007) Effects of ethanol on midbrain neurons: role of opioid receptors. *Alcohol Clin Exp Res* **31**:1106–1113.

Yim CY and Mogenson GJ (1980) Electrophysiological studies of neurons in the ventral tegmental area of Tsai. *Brain Res* **181**:301–313.

Yoshimoto K, McBride WJ, Lumeng L and Li TK (1992) Alcohol stimulates the release of dopamine and serotonin in the nucleus accumbens. *Alcohol* **9**:17–22.

Zhu PJ and Lovinger DM (2006) Ethanol potentiates GABAergic synaptic transmission in a postsynaptic neuron/synaptic bouton preparation from basolateral amygdala. *J Neurophysiol* **96**:433–441.

Zöllner C and Stein C (2007) Opioids. *Handb Exp Pharmacol* **177**:31–63.

Part III
Pain and Injury

Chapter 7
Mechanisms of Plasticity of Inhibition in Chronic Pain Conditions

Charalampos Labrakakis, Francesco Ferrini, and Yves De Koninck

Abstract The balance between inhibition and excitation in the dorsal spinal cord plays a critical role in ensuring that sensory information is relayed accurately to the brain. In particular, a loss of inhibitory control, and the ensuing increase in excitability in spinal dorsal horn neuronal circuits, appears to be a key substrate of pain hypersensitivity. In this Chapter, we summarize the most current knowledge on the involvement of altered GABA and glycine-mediated inhibition in pathological pain. Particular emphasis has been given to the recent finding that altered intracellular chloride homeostasis in neurons of the superficial dorsal horn may explain how inhibition is impaired following peripheral nerve injury and how this may underlie the development of neuropathic pain syndromes. Of particular interest is the finding that this mechanism of injury-induced central disinhibition results from a neuro-immune interaction involving a neuron-to-microglia-to-neuron signalling cascade.

Tissue injury is detected by the organism with the sensation of pain, which triggers defensive responses aiming to terminate the painful experience and protect the organism. Hence, the ability to modulate the pain sensitivity of the tissue surrounding the injury is necessary to protect the damaged tissue and accelerate the healing processes. In other cases however, it is beneficial to reduce the pain felt, as a protective adaptation. For example, lowering pain in a wounded animal fleeing from a predator might increase the possibility of escape. These examples are indicative of the effect of the surrounding conditions on the intensity with which pain is felt and characteristic of dynamic changes in the processing of painful signals. However, during pathological pain conditions, like nerve damage or inflammation, aberrant function of plasticity in the CNS can lead to increased pain sensitivity, that persists over the long term without being directly related to tissue damage. Although such plastic changes in pain sensitivity can occur at several levels of the nociceptive/pain

---

C. Labrakakis (✉)
Division Cellular neurobiology, Centre de recherche université Laval Robert Giffard, Quebec, Canada
and
Department of Biological Applications and Technology, University of Ioannina, Ioannina, Greece
e-mail: c.labrakakis@crulrg.ulaval.ca

M.A. Woodin and A. Maffei (eds.), *Inhibitory Synaptic Plasticity*,
DOI 10.1007/978-1-4419-6978-1_7, © Springer Science+Business Media, LLC 2011

pathway, this chapter will focus on mechanisms originating in the dorsal horn of the spinal cord. The dorsal horn is the first entry point of the nociceptive signal in the CNS and serves as a relay station to higher brain centers. A complex network of interneurons together with afferent nerve terminals, descending control fibers and projection neurons constitute the anatomical substrate for nociceptive information processing. Dynamic changes in the balance between the excitatory and inhibitory neurotransmission in the dorsal horn will influence the outcome of nociceptive information and may lead to aberrant pain sensation.

Past work has focused mostly on the plasticity of excitatory transmission in the effort to elucidate the mechanism of pathological pain. Recent work however is indicating that plasticity of inhibitory transmission is an equally crucial factor. The significance of spinal inhibitory transmission in the development of pathologic pain has been demonstrated in many paradigms. Block of GABA and glycine inhibition in the spinal cord promotes the transmission of nociceptive information (hyperalgesia) and makes innocuous input painful (allodynia), akin to what is observed in neuropathic and inflammatory pain (Yaksh 1989; Sivilotti and Woolf 1994; Sherman and Loomis 1994; Sorkin and Puig 1996; Sherman et al. 1997). At the cellular level, block of inhibition unmasks a network of low-threshold polysynaptic input onto superficial dorsal horn neurons (Baba et al. 2003; Torsney and Macdermott 2006). Comparably, nociceptive specific lamina I neurons show responses to innocuous touch after depression of intrinsic $GABA_A$/glycine inhibition (Keller et al. 2007) and this normally innocuous information is relayed through ascending pathways to nociceptive specific thalamic neurons (Sherman et al. 1997) (Fig. 7.1).

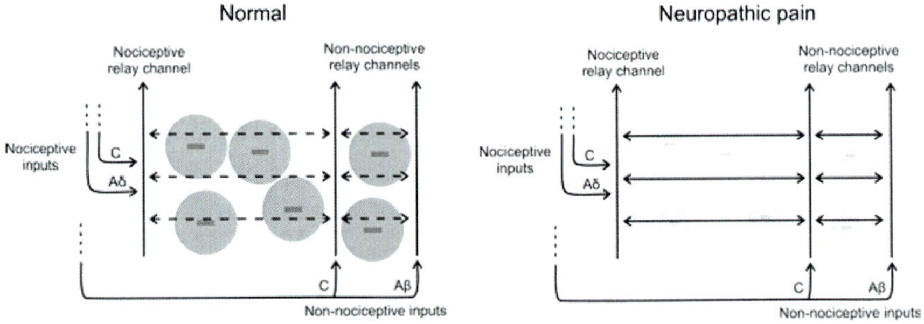

Fig. 7.1 Summary diagram illustrating how loss of inhibition at the spinal level can unmask low-threshold input to nociceptive relay channels in the dorsal horn of the spinal cord to explain allodynia. Normally, low-threshold (innocuous) input is conveyed through Aβ mechanosensitive afferents and C innocuous thermal or brush afferents to non-nociceptive relay neurons. In contrast, high-threshold (nociceptive) input is conveyed through Aδ and C mechanosensitive and thermosensitive afferents. While these afferent connections terminate on separate populations of neurons, yielding separate relay pathways, the pathways are interconnected via a circuit of local interneurons. Normally, local spinal inhibition (*dark grey circles*) maintains this interconnection silent and thus these pathways separated (*left panel*). However, disinhibition (*light grey circles*) can unmask these interconnections, allowing for a crosstalk between pathways and, for example, relay of innocuous input via normally nociceptive specific pathways, providing a substrate for the perception of pain in response to a normally innocuous stimulus (allodynia; *right panel*) (Baba et al. 2003; Torsney and Macdermott 2006; Keller et al. 2007)

As both nerve injury and peripheral inflammation appear to lead to a loss of functional inhibition, identification of the intrinsic mechanisms of inhibitory plasticity involved in neuropathic and inflammatory pain might provide improved opportunities for target-specific treatments. Such mechanisms might involve changes at the presynaptic level or at the postsynaptic target of inhibition.

Plasticity of Spinal Inhibition by Altering the Source and Release of GABA and Glycine

At the presynaptic level, plastic changes have as a final result the modulation of the amount of neurotransmitter reaching the postsynaptic neurons. Several mechanisms that involve presynaptic changes causing a decrease in available inhibitory neurotransmitters have been investigated.

Loss of Inhibitory Neurons as a Cause of Central Pain Hypersensitivity

A reduction of GABAergic cells in the dorsal horn and a decrease of GABA has been observed after sciatic nerve transection (Castro-Lopes et al. 1993). In addition, several studies describe apoptosis or degeneration of inhibitory interneurons following peripheral nerve lesion (Ibuki et al. 1997; Moore et al. 2002; Scholz et al. 2005), and accordingly local use of antiapoptotic drugs attenuates pain behaviour (Scholz et al. 2005). However, by using a stereological approach to quantify cell death, other reports have provided evidence that certain nerve lesions cause pain hypersensitivity in the absence of significant apoptosis in the dorsal horn (Polgar et al. 2004; Polgar et al. 2005). The question therefore remains of whether pain hypersensitivity after nerve lesion is necessarily linked to apoptosis of inhibitory interneurons.

Pain as a Result of Changes in GABA Synthesis

Irrespectively of the absolute number of interneurons, a significant reduction of inhibition can be due to a reduced synthesis of GABA (Eaton et al. 1998; Somers and Clemente 2002). In neuropathic animal models, a loss of GABA content in the dorsal horn synapses has been reported (Somers and Clemente 2002) and this process seems to be associated with a down-regulation of the GABA synthesizing enzyme glutamic acid decarboxylase (GAD; Eaton et al. 1998). Interestingly, GAD expression is rather increased in inflammatory pain models (Castro-Lopes et al. 1993, 1994).

Modulation of GABA/Glycine Release from Presynaptic Terminals

The release of GABA/glycine can also be regulated by specific presynaptic receptors. In particular, $GABA_B$ and glutamate receptors are expressed on presynaptic inhibitory terminals and their activation can modulate the transmitter release (Chéry and De Koninck 2000; Kerchner et al. 2001; Hugel and Schlichter 2003; Lu et al. 2005; Engelman et al. 2006; Choi et al. 2008). Direct effects on inhibitory transmitter release can also be induced by the activation of serotonin receptors (Fukushima et al. 2009), muscarinic receptors (Baba et al. 1998; Wang et al. 2006), nicotinic receptors (Kiyosawa et al. 2001; Cordero-Erausquin et al. 2004), purinergic receptors (Jang et al. 2001; Hugel and Schlichter 2003), adrenergic receptors (Gassner et al. 2009), neurokinin receptors (Vergnano et al. 2004), and TrkB receptors (Bardoni et al. 2007).

Recently a novel spinal mechanism of GABAergic presynaptic plasticity mediated by endocannabinoids has been suggested (Pernia-Andrade et al. 2009). Cannabinoids are known for the analgesic action that they exert by binding to CB1 and CB2 receptors on peripheral nociceptors (Walker and Hohmann 2005; Agarwal et al. 2007). In contrast, in the dorsal horn, cannabinoids have been found to mediate pro-nociceptive actions (Pernia-Andrade et al. 2009). CB1 receptors are densely expressed in the superficial dorsal horn (Farquhar-Smith et al. 2000) and are localized presynaptically on inhibitory terminals. Their activation reduces the synaptic release of glycine and GABA, hence reducing inhibition and consequently causing disinhibition. The source of cannabinnoids are local excitatory neurons. Their depolarization by intense stimuli, causes the production of endocannabinoids which then can act retrogradely to weaken the inhibitory input they receive. This activity dependent pro-nociceptive effect could explain the secondary mechanical hyperalgesia triggered by intense nociceptive input in the surrounding skin.

Plasticity of Spinal Inhibition at the Postsynaptic Level

At the postsynaptic level the traditional targets for plastic modulation are the inhibitory neurotransmitter receptors. In addition, recent work has revealed a new mechanism of inhibitory synaptic strength control by modulating the transmembrane anion gradient:

Modulation of Spinal Postsynaptic Glycine Receptors by Inflammatory Mediators

Increased pain sensitivity during tissue inflammation is partly mediated by the production and release of various prostaglandins. Their role however, is not limited in the peripheral tissue where the inflammation occurs, but it also extends centrally, mediating central sensitization. Cyclooxygenase 2, the rate limiting enzyme of the prostaglandin production pathway, is upregulated in the spinal dorsal horn after tissue inflammation, leading to increased production of prostaglandin E2 (Guhring et al. 2000; Samad et al. 2001). PGE2 selectively blocks glycinergic transmission in the dorsal horn, by phosphorylating the α3 glycine receptor subunit (Ahmadi et al. 2002). Accordingly, mice deficient for the α3 subunit show reduced PGE2 and inflammation mediated pain sensitization (Harvey et al. 2004). Interestingly, this mechanism does not contribute to pain after nerve injury (Hosl et al. 2006). The significance of these findings lies in the potential development of target specific pharmacological treatments that have reduced side effects.

Modulation of Spinal $GABA_A$ Receptor Function by Endogenous Neurosteroids After Inflammation

A form of postsynaptic receptor modulation of $GABA_A$ receptor signalling is mediated by neurosteroids, which cause a prolongation of synaptic inhibition (Harrison et al. 1987; Cooper et al. 1999). While in the developing dorsal horn endogenous 5α-reduced neurosteroids are tonically released, in the adult dorsal horn their production ceases. This developmental change results in an acceleration of GABAergic current kinetics in adults (Keller et al. 2004). Yet, after peripheral tissue inflammation neurosteroid production in adult dorsal horn is reactivated again, which causes a deceleration of GABA receptor kinetics (Poisbeau et al. 2005). Unlike the previous examples of plasticity, in which plastic changes during pathological pain were leading to disinhibition and a pronociceptive behavior, this mechanism leads to increased inhibition and thus is antinociceptive. Indeed, the increased inhibition correlated with a decrease in inflammation-induced thermal hyperalgesia by endogenous neurosteroid production (Poisbeau et al. 2005).

Altered Chloride Homeostasis in Pain

Emerging evidence of the modulation of ion gradients in adult tissue represents an intriguing mechanism to control synaptic strength, and therefore provides a novel perspective on synaptic plasticity. In particular, evidence of altered chloride homeostasis in pathological conditions also offers novel avenues for therapeutic interventions (De Koninck 2007). Below, we will review in details the cascade of mechanisms leading to altered chloride homeostasis in the spinal dorsal horn following peripheral nerve injury and the functional impact of such phenomenon.

Intracellular Chloride Concentration Defines the Polarity of GABA/Glycine-Mediated Currents

$GABA_A$ and glycine receptors are ligand-gated ion channels permeable to chloride (Cl^-) and bicarbonate ($HCO3^-$). The concentration of these anions in the intracellular and extracellular media sets the reversal potential for GABA currents (E_{GABA}), as well as for glycine ($E_{glycine}$), therefore determining their overall effect on the response properties of the cells to synaptic inputs. Since Cl^- is a small largely available anion and its permeability is about fourfold the one for $HCO3^-$, $E_{GABA/glycine}$ mostly depends on Cl^- concentration. In addition, the intracellular Cl^- concentration is much lower than the extracellular one (5–20-fold), so that small or millimolar changes in the intracellular concentration may strongly challenge the transmembrane gradient. Under physiological conditions in mature neurons the intracellular Cl^- concentration is relatively low (less than 10 mM) and the $E_{GABA/glycine}$ is more negative than the cell resting potential (Vr). Consequently, the opening of GABA/glycine receptors leads to a Cl^- influx that in turn hyperpolarizes the cell membrane and reduces the cell excitability. However, a strong hyperpolarization could lead to rebound action potentials in some cases (Bevan et al. 2002; Baufreton et al. 2005). On the other hand a small depolarization could cause shunting inhibition by decreasing the membrane resistance and affecting the membrane time constant (Mitchell and Silver 2003; Prescott et al. 2006). In this respect, the equations "inhibition" = "hyperpolarization" and "excitation" = "depolarization" are not practical and therefore, a better way to define GABA/glycine "inhibition" should be sought in the net effect produced on neuronal firing.

KCC2 and Chloride Homeostasis in CNS

In mature neurons, the intracellular Cl^- concentration is kept low by a potassium-chloride co-transporter known as KCC2. KCC2 belongs to the family of the SLC12 cation-chloride co-transporters (CCCs), a group of glycoproteins involved in the electroneutral transport of ions across the cell membranes (Payne et al. 2003; Gamba 2005; Price et al. 2005). Together with the sodium-potassium-chloride co-transporter-1 (NKCC1), they are considered to be the major players in maintaining Cl^- balance in the nervous system (Payne et al. 2003). While the first is the main Cl^- extruder in neurons, the latter facilitates Cl^- accumulation. However, their temporal and spatial localizations seem to be in general different. At early developmental stages, NKCC1 expression levels are high while KCC2 levels are low (consequently, due to the high intracellular Cl^- concentration, GABA is typically depolarizing in immature neurons). With maturation, NKCC1 expression diminishes, and KCC2 expression increases, causing a shift in Cl^- ion gradients. On the other hand, high levels of NKCC1 appear to be maintained in most of the mature neurons in the peripheral nervous system and in some specific sub-cellular compartments within the central nervous system (Khirug et al. 2008).

Since intracellular Cl^- concentration depends on KCC2, any change in its activity and/or expression may affect the strength or even the polarity of GABAergic/glycinergic neurotransmission. To date, an increasing number of studies have shown that the regulation of KCC2 activity represents a key mechanism for disinhibition in the adult nervous system under both physiological (Hewitt et al. 2009) and pathological (Rivera et al. 2002; Jin et al. 2005; Palma et al. 2006) conditions.

Altered KCC2 Activity in the Spinal Dorsal Horn After Nerve Lesion

In the spinal dorsal horn, the down-regulation of KCC2 activity specifically affects post-synaptic inhibition of projection neurons in neuropathic animals (Coull et al. 2003). Indeed, after peripheral nerve injury, E_{GABA} of spinal lamina I neurons is shifted toward more positive values significantly reducing the hyperpolarizing effect of GABA and in some cases converting inhibition to a net excitation. This appears to be due to a loss of KCC2 activity, since KCC2 expression is reduced in different models of neuropathic pain (Coull et al. 2003; Lu et al. 2008; Jolivalt et al. 2008; Miletic and Miletic 2008; Cramer et al. 2008). Loss of KCC2 activity also seems to be a sufficient mechanism to explain neuropathic pain symptoms: local spinal blockade or knock-down of spinal KCC2 lowers the nociceptive withdrawal threshold and mimics the symptoms of neuropathic pain (Coull et al. 2003) and changes the phenotype of lamina I neurons in control animals by unmasking innocuous mechanical inputs to these normally nociceptive specific neurons (Keller et al. 2007). The latter mechanism thus appears sufficient to explain how innocuous input can provoke a nociceptive response (allodynia).

From Injured Primary Afferent to KCC2 Downregulation in the Dorsal Horn Neurons

KCC2 function can be regulated by several mechanisms. In particular, the transporter activity has been suggested to be modulated by both protein oligomerization (Blaesse et al. 2006) and phosphorylation states (Khirug et al. 2005; Rinehart et al. 2009; Watanabe et al. 2009). Indeed, an increased level of oligomerization observed during the development seems to be correlated with an increased level of KCC2 activity (Blaesse et al. 2006). With regards to expression of KCC2 in neurons, Rivera and colleagues (Rivera et al. 2002; Rivera et al. 2004) have shown that it is regulated via the BDNF-TrkB signalling pathway. Indeed, they showed that exposing hippocampal cultures to BDNF caused a rapid (within less than 1 h) down-regulation of KCC2 mRNA and protein in neurons (Rivera et al. 2002). In addition, they identified two major second messenger pathways coupled to TrkB receptors that control KCC2 expression: the Shc/FRS-2 (src homology 2 domain-containing transforming protein/fibroblast growth factor receptor substrate 2) pathway and the PLCγ/CREB (phospholipase C-γ/cyclic adenosine monophosphate response element-binding protein) pathway (Rivera et al. 2004). Interestingly, activation of both pathways leads to a down-regulation of KCC2, whereas activation of Shc pathway alone induce an up-regulation of KCC2, suggesting a bi-directional action of BDNF that is consistent with the role of BDNF in the up-regulation of KCC2 during the early postnatal period (Aguado et al. 2003). This modulatory role of BDNF on KCC2 fits well with its known role as a pain modulator (Merighi et al. 2008). Indeed, intrathecal administration of BDNF significantly increases pain sensitivity in rats and its acute application on spinal cord slices induces a depolarizing shift on E_{GABA} (Coull et al. 2005). Conversely, removing endogenous BDNF by using TrkB-Fc fusion protein or TrkB blocking antibody acutely reverses the depolarizing shift on E_{GABA} observed in lamina I neurons of neuropathic animals (Coull et al. 2005). Therefore, plastic changes in spinal inhibition induced by peripheral nerve injury are mediated through a signalling pathway that involves the release of BDNF, the activation of TrkB receptors and the down-regulation of KCC2, with consequent disruption of Cl^- homeostasis.

Activated Microglia Trigger Down-Regulation of KCC2 by Releasing BDNF

An important source of BDNF in the spinal cord following injury appears to be microglia, the resident macrophages in the central nervous system. After spinal injury or nerve lesion (and many other different insults directed against the central nervous system) microglia acquire a new morphological and biochemical reactive profile and release pro-inflammatory or cell-signalling molecules. Even though several pathways can lead to this activated state, the purinergic system associated with microglia has been shown to play a pivotal role in the physiopathology of neuropathic pain (Trang et al. 2006; Inoue 2008). ATP released by injured neurons, and possibly glial cells, promotes microglial activation and chemotaxis (Honda et al. 2001). Microglia activation involves the expression *de novo* of the purinergic $P2X_4$ receptors (Tsuda et al. 2003; Ulmann et al. 2008; Trang et al. 2009) and this process plays a major role in the development of neuropathic pain behaviour. Indeed, intrathecal administration of antisense oligodeoxynucleotide targeting $P2X_4$ reduces tactile allodynia in nerve injury models, whereas administration of ATP-stimulated microglia elicits allodynia in naïve animals (Tsuda et al. 2003). $P2X_4$ null mice also do not develop pain hypersensitivity after peripheral nerve injury (Ulmann et al. 2008). Interestingly, ATP-stimulated microglia release BDNF in a $P2X_4$-dependent manner (Ulmann et al. 2008; Trang et al. 2009) and intrathecal administration of activated microglia that have been pretreated with siRNA against BDNF does not elicit allodynia (Coull et al. 2005). Taken together, these findings indicate that ATP-stimulated microglia, by releasing BDNF, activates the TrkB mediated intracellular pathways described above, thus affecting the KCC2 activity. In summary, after nerve injury a series of events develop, starting with the activation of spinal microglia and ending with the alteration of Cl^- extrusion capacity of lamina I neurons (Coull et al. 2005) (Fig. 7.2). Importantly, local blockade of $P2X_4$ in spinal slices extracted from nerve injured animals causes a return of E_{GABA} toward more negative values (Coull et al. 2005), suggesting that the microglia-neuronal signalling is continuously maintained in neuropathic animals and its effect can be reversed.

Yet, it is not easy to predict the functional impact of a change in Cl^- homeostasis on cells and network excitability. Using NEURON models, where E_{GABA} could be adjusted artificially, we found that even very small sub-threshold shifts in E_{GABA} (by as little as 5 mV) were sufficient to significantly affect the input-output properties

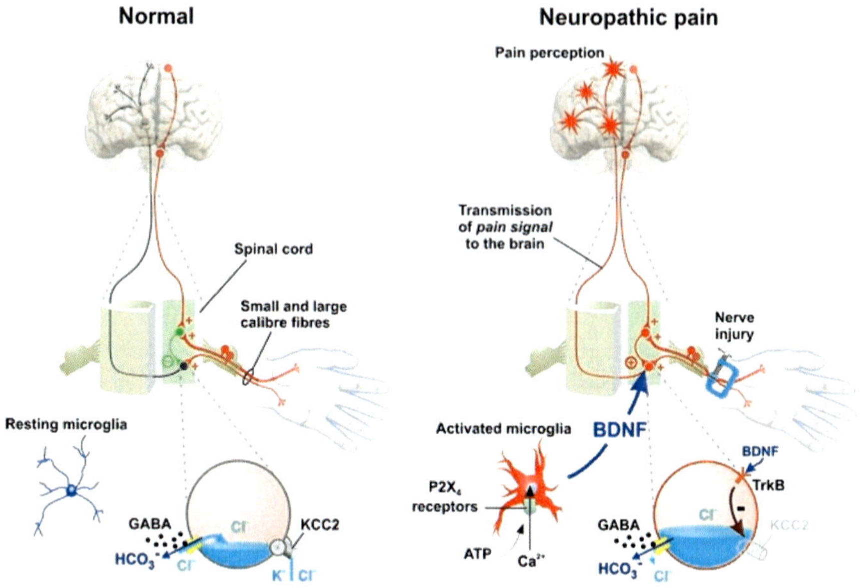

Fig. 7.2 Summary diagram illustrating the cascade of events leading to spinal disinhibition following peripheral nerve injury and the relay of innocuous input to normally nociceptive output neurons. Under normal conditions, intracellular Cl^- concentration is sufficiently low that when GABA or glycine opens $GABA_A$/glycine receptor-channels, a net outward current results, causing hyperpolarizing inhibition. After nerve injury, resting microglia becomes activated and begins to express the $P2X_4$ receptor (Tsuda et al. 2003). When ATP binds to these $P2X_4$ receptors, the activated microglia release BDNF which acts on TrkB receptors on dorsal horn neurons causing a downregulation of KCC2 and accumulation of intracellular Cl^- (Rivera et al. 2002; Rivera et al. 2004; Coull et al. 2005). The net result is a collapse of outward, hyperpolarizing, Cl^- current either decreasing net hyperpolarizing inhibition or even converting the current to a depolarizing one because of the dominant inward HCO_3^- current (Cordero-Erausquin et al. 2005). This leads to a loss of inhibition at the spinal level, and in some extreme cases, the net conversion of inhibition into excitation (Coull et al. 2003). Innocuous input which is normally not communicated to nociceptive output neurons is now unmasked and triggers the aberrant relay of a nociceptive signal to the brain where pain is perceived (Keller et al. 2007)

of neurons (Prescott et al. 2006). To fully appreciate the impact of altered Cl^- homeostasis on how neurons integrate synaptic inputs, one must also consider the spatiotemporal dynamics of Cl^- during an input. This is well illustrated by findings we made from studying the early stages of spinal cord development in rats: During the first postnatal week, GABA is depolarizing in dorsal horn neurons, consistent with increased nociceptive responses in young rats. After the first post-natal week, GABA action switches to hyperpolarizing responses (Baccei and Fitzgerald 2004), however nociceptive responses do not reach mature levels until the third post-natal week. Closer investigation showed that, during this period, GABA applications produced biphasic responses, consisting in an initial hyperpolarisation, followed by a depolarization. This temporal switch was a consequence of intracellular Cl^- accumulation due to Cl^- inflow through $GABA_A$ receptors, unmasking the bicarbonate-mediated depolarizing component (Cordero-Erausquin et al. 2005). Such a biphasic action of GABA could occur during repetitive $GABA_A$-mediated synaptic inputs indicating a weaker capacity of the young spinal dorsal horn neurons (7–14 postnatal days) to maintain a low intracellular Cl^- concentration when challenged by sustained inhibitory input. This dynamic collapse of inhibition during repetitive $GABA_A$-input can likely explain the lower nociceptive withdrawal threshold of young rats (Cordero-Erausquin et al. 2005).

Conclusion

Chronic and neuropathic pain induces plastic changes in synaptic transmission at the dorsal horn level, and the main consequence of these changes is an increased sensitivity to pain stimuli or even spontaneous pain in absence of stimuli. In the present chapter, we have discussed the most recent experimental studies showing that a large part of these plastic changes consists in a loss of inhibition. In particular, we have focused on the recent findings that impaired Cl^- extrusion appear to be a significant substrate of such disinhibition. One of the difficulties in treating pathological pain conditions resides in the lack of drugs that can effectively restore the physiological inhibitory tone. Recent research outlining the importance of specific modulatory mechanisms of $GABA_A$ and glycine receptors in the development of inflammatory and neuropathic pain may provide more specific and selective approaches to control hyperexcitability of sensory neurons. In particular, the finding that the $\alpha 3$ subunits of the glycine receptor are selective targets of prostaglandin E2 provides guidance for the development of novel therapeutic drugs (Zeilhofer and Brune 2006). Similarly, the findings that benzodiazepines that do not act on the $\alpha 1$ subunit of the $GABA_A$ receptor can produce analgesia without sedation offer promising avenues for the development of novel analgesics with fewer side effects (Zeilhofer et al. 2009). However, the finding of altered Cl^- homeostasis further complicates this strategy. Indeed, while some drugs such as benzodiazepines can effectively produce analgesia by enhancing spinal inhibitory neurotransmission (Knabl et al. 2008), the same drugs may be less effective, or lead to paradoxical hyperalgesia, if an alteration of Cl^- homeostasis occurs. In this context, it is interesting to note that a recent study showed that while benzodiazepines are analgesic at lower doses in animals with nerve injury, these drugs become counterproductive at higher doses (Asiedu et al. 2009), consistent with the prediction that with reduced Cl^- extrusion capacity, enhancing $GABA_A$ responses will eventually lead to rebound excitation (Cordero-Erausquin et al. 2005).

Acknowledgements The authors acknowledge support by the Canadian Institutes of Health Research (CIHR; grants to YDeK). C. Labrakakis was supported by a post-doctoral Fellowship from CIHR. YDeK is a *Chercheur National* of the Fonds de la recherche en santé du Québec. We thank Mr. Sylvain Côté for expert assistance in preparing the illustrations.

References

Agarwal N, Pacher P, Tegeder I, Amaya F, Constantin CE, Brenner GJ, Rubino T, Michalski CW, Marsicano G, Monory K, Mackie K, Marian C, Batkai S, Parolaro D, Fischer MJ, Reeh P, Kunos G, Kress M, Lutz B, Woolf CJ, Kuner R (2007) Cannabinoids mediate analgesia largely via peripheral type 1 cannabinoid receptors in nociceptors. Nat Neurosci 10:870–879

Aguado F, Carmona MA, Pozas E, Aguilo A, Martinez-Guijarro FJ, Alcantara S, Borrell V, Yuste R, Ibanez CF, Soriano E (2003) BDNF regulates spontaneous correlated activity at early developmental stages by increasing synaptogenesis and expression of the K^+/Cl^- co-transporter KCC2. Development 130:1267–1280

Ahmadi S, Lippross S, Neuhuber WL, Zeilhofer HU (2002) PGE(2) selectively blocks inhibitory glycinergic neurotransmission onto rat superficial dorsal horn neurons. Nat Neurosci 5:34–40

Asiedu M, Ossipov MH, Kaila K, Price TJ (2009) Acetazolamide and midazolam act synergistically to inhibit neuropathic pain. Pain

Baba H, Ji RR, Kohno T, Moore KA, Ataka T, Wakai A, Okamoto M, Woolf CJ (2003) Removal of GABAergic inhibition facilitates polysynaptic A fiber-mediated excitatory transmission to the superficial spinal dorsal horn. Mol Cell Neurosci 24:818–830

Baba H, Kohno T, Okamoto M, Goldstein PA, Shimoji K, Yoshimura M (1998) Muscarinic facilitation of GABA release in substantia gelatinosa of the rat spinal dorsal horn. J Physiol 508 (Pt 1):83–93

Baccei ML, Fitzgerald M (2004) Development of GABAergic and glycinergic transmission in the neonatal rat dorsal horn. J Neurosci 24:4749–4757

Bardoni R, Ghirri A, Salio C, Prandini M, Merighi A (2007) BDNF-mediated modulation of GABA and glycine release in dorsal horn lamina II from postnatal rats. Dev Neurobiol 67:960–975

Baufreton J, Atherton JF, Surmeier DJ, Bevan MD (2005) Enhancement of excitatory synaptic integration by GABAergic inhibition in the subthalamic nucleus. J Neurosci 25:8505–8517

Bevan MD, Magill PJ, Hallworth NE, Bolam JP, Wilson CJ (2002) Regulation of the timing and pattern of action potential generation in rat subthalamic neurons in vitro by GABA-A IPSPs. J Neurophysiol 87:1348–1362

Blaesse P, Guillemin I, Schindler J, Schweizer M, Delpire E, Khiroug L, Friauf E, Nothwang HG (2006) Oligomerization of KCC2 correlates with development of inhibitory neurotransmission. J Neurosci 26:10407–10419

Castro-Lopes JM, Tavares I, Coimbra A (1993) GABA decreases in the spinal cord dorsal horn after peripheral neurectomy. Brain Res 620:287–291

Castro-Lopes JM, Tölle TR, Pan B, Zieglgänsberger W (1994) Expression of GAD mRNA in spinal cord neurons of normal and monoarthritic rats. Brain Res Mol Brain Res 26:169–176

Chéry N, De Koninck Y (2000) $GABA_B$ receptors are the first target of GABA released at lamina I inhibitory synapses in the rat spinal cord. J Neurophysiol 84:1006–1011

Choi IS, Cho JH, Jeong SG, Hong JS, Kim SJ, Kim J, Lee MG, Choi BJ, Jang IS (2008) GABA(B) receptor-mediated presynaptic inhibition of glycinergic transmission onto substantia gelatinosa neurons in the rat spinal cord. Pain 138:330–342

Cooper EJ, Johnston GA, Edwards FA (1999) Effects of a naturally occurring neurosteroid on GABAA IPSCs during development in rat hippocampal or cerebellar slices. J Physiol 521 Pt 2:437–449

Cordero-Erausquin M, Coull JA, Boudreau D, Rolland M, De Koninck Y (2005) Differential maturation of GABA action and anion reversal potential in spinal lamina I neurons; impact of chloride extrusion capacity. J Neurosci 25:9613–9623

Cordero-Erausquin M, Pons S, Faure P, Changeux JP (2004) Nicotine differentially activates inhibitory and excitatory neurons in the dorsal spinal cord. Pain 109:308–318

Coull JA, Beggs S, Boudreau D, Boivin D, Tsuda M, Inoue K, Gravel C, Salter MW, De Koninck Y (2005) BDNF from microglia causes the shift in neuronal anion gradient underlying neuropathic pain. Nature 438:1017–1021

Coull JA, Boudreau D, Bachand K, Prescott SA, Nault F, Sik A, De Koninck P, De Koninck Y (2003) Trans-synaptic shift in anion gradient in spinal lamina I neurons as a mechanism of neuropathic pain. Nature 424:938–942

Cramer SW, Baggott C, Cain J, Tilghman J, Allcock B, Miranpuri G, Rajpal S, Sun D, Resnick D (2008) The role of cation-dependent chloride transporters in neuropathic pain following spinal cord injury. Mol Pain 4:36

De Koninck Y (2007) Altered chloride homeostasis in neurological disorders: a new target. Curr Opin Pharmacol 7:93–99

Eaton MJ, Plunkett JA, Karmally S, Martinez MA, Montanez K (1998) Changes in GAD- and GABA- immunoreactivity in the spinal dorsal horn after peripheral nerve injury and promotion of recovery by lumbar transplant of immortalized serotonergic precursors. J Chem Neuroanat 16:57–72

Engelman HS, Anderson RL, Daniele C, Macdermott AB (2006) Presynaptic alpha-amino-3-hydroxy-5-methyl-4-isoxazolepropionic acid (AMPA) receptors modulate release of inhibitory amino acids in rat spinal cord dorsal horn. Neuroscience 139:539–553

Farquhar-Smith WP, Egertova M, Bradbury EJ, McMahon SB, Rice AS, Elphick MR (2000) Cannabinoid CB(1) receptor expression in rat spinal cord. Mol Cell Neurosci 15:510–521

Fukushima T, Ohtsubo T, Tsuda M, Yanagawa Y, Hori Y (2009) Facilitatory actions of serotonin type 3 receptors on GABAergic inhibitory synaptic transmission in the spinal superficial dorsal horn. J Neurophysiol 102:1459–1471

Gamba G (2005) Molecular physiology and pathophysiology of electroneutral cation-chloride cotransporters. Physiol Rev 85:423–493

Gassner M, Ruscheweyh R, Sandkuhler J (2009) Direct excitation of spinal GABAergic interneurons by noradrenaline. Pain 145:204–210

Guhring H, Gorig M, Ates M, Coste O, Zeilhofer HU, Pahl A, Rehse K, Brune K (2000) Suppressed injury-induced rise in spinal prostaglandin E2 production and reduced early thermal hyperalgesia in iNOS-deficient mice. J Neurosci 20:6714–6720

Harrison NL, Vicini S, Barker JL (1987) A steroid anesthetic prolongs inhibitory postsynaptic currents in cultured rat hippocampal neurons. J Neurosci 7:604–609

Harvey RJ, Depner UB, Wassle H, Ahmadi S, Heindl C, Reinold H, Smart TG, Harvey K, Schutz B, Abo-Salem OM, Zimmer A, Poisbeau P, Welzl H, Wolfer DP, Betz H, Zeilhofer HU, Muller U (2004) GlyR alpha3: an essential target for spinal PGE2-mediated inflammatory pain sensitization. Science 304:884–887

Hewitt SA, Wamsteeker JI, Kurz EU, Bains JS (2009) Altered chloride homeostasis removes synaptic inhibitory constraint of the stress axis. Nat Neurosci 12:438–443

Honda S, Sasaki Y, Ohsawa K, Imai Y, Nakamura Y, Inoue K, Kohsaka S (2001) Extracellular ATP or ADP induce chemotaxis of cultured microglia through Gi/o-coupled P2Y receptors. J Neurosci 21:1975–1982
Hosl K, Reinold H, Harvey RJ, Muller U, Narumiya S, Zeilhofer HU (2006) Spinal prostaglandin E receptors of the EP2 subtype and the glycine receptor alpha3 subunit, which mediate central inflammatory hyperalgesia, do not contribute to pain after peripheral nerve injury or formalin injection. Pain 126:46–53
Hugel S, Schlichter R (2003) Convergent control of synaptic GABA release from rat dorsal horn neurones by adenosine and GABA autoreceptors. J Physiol 551:479–489
Ibuki T, Hama AT, Wang XT, Pappas GD, Sagen J (1997) Loss of GABA-immunoreactivity in the spinal dorsal horn of rats with peripheral nerve injury and promotion of recovery by adrenal medullary grafts. Neuroscience 76:845–858
Inoue K (2008) Purinergic systems in microglia. Cell Mol Life Sci 65:3074–3080
Jang IS, Rhee JS, Kubota H, Akaike N, Akaike N (2001) Developmental changes in P2X purinoceptors on glycinergic presynaptic nerve terminals projecting to rat substantia gelatinosa neurones. J Physiol 536:505–519
Jin X, Huguenard JR, Prince DA (2005) Impaired Cl^- extrusion in layer V pyramidal neurons of chronically injured epileptogenic neocortex. J Neurophysiol 93:2117–2126
Jolivalt CG, Lee CA, Ramos KM, Calcutt NA (2008) Allodynia and hyperalgesia in diabetic rats are mediated by GABA and depletion of spinal potassium-chloride co-transporters. Pain 140:48–57
Keller AF, Beggs S, Salter MW, De Koninck Y (2007) Transformation of the output of spinal lamina I neurons after nerve injury and microglia stimulation underlying neuropathic pain. Mol Pain 3:27
Keller AF, Breton JD, Schlichter R, Poisbeau P (2004) Production of 5alpha-reduced neurosteroids is developmentally regulated and shapes GABA(A) miniature IPSCs in lamina II of the spinal cord. J Neurosci 24:907–915
Kerchner GA, Wang GD, Qiu CS, Huettner JE, Zhuo M (2001) Direct presynaptic regulation of GABA/glycine release by kainate receptors in the dorsal horn: an ionotropic mechanism. Neuron 32:477–488
Khirug S, Huttu K, Ludwig A, Smirnov S, Voipio J, Rivera C, Kaila K, Khiroug L (2005) Distinct properties of functional KCC2 expression in immature mouse hippocampal neurons in culture and in acute slices. Eur J Neurosci 21:899–904
Khirug S, Yamada J, Afzalov R, Voipio J, Khiroug L, Kaila K (2008) GABAergic depolarization of the axon initial segment in cortical principal neurons is caused by the Na-K-2Cl cotransporter NKCC1. J Neurosci 28:4635–4639
Kiyosawa A, Katsurabayashi S, Akaike N, Pang ZP, Akaike N (2001) Nicotine facilitates glycine release in the rat spinal dorsal horn. J Physiol 536:101–110
Knabl J, Witschi R, Hosl K, Reinold H, Zeilhofer UB, Ahmadi S, Brockhaus J, Sergejeva M, Hess A, Brune K, Fritschy JM, Rudolph U, Mohler H, Zeilhofer HU (2008) Reversal of pathological pain through specific spinal GABAA receptor subtypes. Nature 451:330–334
Lu CR, Willcockson HH, Phend KD, Lucifora S, Darstein M, Valtschanoff JG, Rustioni A (2005) Ionotropic glutamate receptors are expressed in GABAergic terminals in the rat superficial dorsal horn. J Comp Neurol 486:169–178
Lu Y, Zheng J, Xiong L, Zimmermann M, Yang J (2008) Spinal cord injury-induced attenuation of GABAergic inhibition in spinal dorsal horn circuits is associated with down-regulation of the chloride transporter KCC2 in rat. J Physiol 586:5701–5715
Merighi A, Salio C, Ghirri A, Lossi L, Ferrini F, Betelli C, Bardoni R (2008) BDNF as a pain modulator. Prog Neurobiol 85:297–317
Miletic G, Miletic V (2008) Loose ligation of the sciatic nerve is associated with TrkB receptor-dependent decreases in KCC2 protein levels in the ipsilateral spinal dorsal horn. Pain 137:532–539
Mitchell SJ, Silver RA (2003) Shunting inhibition modulates neuronal gain during synaptic excitation. Neuron 38:433–445

Moore KA, Kohno T, Karchewski LA, Scholz J, Baba H, Woolf CJ (2002) Partial peripheral nerve injury promotes a selective loss of GABAergic inhibition in the superficial dorsal horn of the spinal cord. J Neurosci 22:6724–6731

Palma E, Amici M, Sobrero F, Spinelli G, Di Angelantonio S, Ragozzino D, Mascia A, Scoppetta C, Esposito V, Miledi R, Eusebi F (2006) Anomalous levels of Cl^- transporters in the hippocampal subiculum from temporal lobe epilepsy patients make GABA excitatory. Proc Natl Acad Sci U S A 103:8465–8468

Payne JA, Rivera C, Voipio J, Kaila K (2003) Cation-chloride co-transporters in neuronal communication, development and trauma. Trends Neurosci 26:199–206

Pernia-Andrade AJ, Kato A, Witschi R, Nyilas R, Katona I, Freund TF, Watanabe M, Filitz J, Koppert W, Schuttler J, Ji G, Neugebauer V, Marsicano G, Lutz B, Vanegas H, Zeilhofer HU (2009) Spinal endocannabinoids and CB1 receptors mediate C-fiber-induced heterosynaptic pain sensitization. Science 325:760–764

Poisbeau P, Patte-Mensah C, Keller AF, Barrot M, Breton JD, Luis-Delgado OE, Freund-Mercier MJ, Mensah-Nyagan AG, Schlichter R (2005) Inflammatory pain upregulates spinal inhibition via endogenous neurosteroid production. J Neurosci 25:11768–11776

Polgar E, Gray S, Riddell JS, Todd AJ (2004) Lack of evidence for significant neuronal loss in laminae I-III of the spinal dorsal horn of the rat in the chronic constriction injury model. Pain 111:144–150

Polgar E, Hughes DI, Arham AZ, Todd AJ (2005) Loss of neurons from laminas I-III of the spinal dorsal horn is not required for development of tactile allodynia in the spared nerve injury model of neuropathic pain. J Neurosci 25:6658–6666

Prescott SA, Sejnowski TJ, De Koninck Y (2006) Reduction of anion reversal potential subverts the inhibitory control of firing rate in spinal lamina I neurons: a biophysical basis for neuropathic pain. Mol Pain 2:32

Price TJ, Cervero F, De Koninck Y (2005) Role of cation-chloride-cotransporters (CCC) in pain and hyperalgesia. Curr Top Med Chem 5:547–555

Rinehart J, Maksimova YD, Tanis JE, Stone KL, Hodson CA, Zhang J, Risinger M, Pan W, Wu D, Colangelo CM, Forbush B, Joiner CH, Gulcicek EE, Gallagher PG, Lifton RP (2009) Sites of regulated phosphorylation that control K-Cl cotransporter activity. Cell 138:525–536

Rivera C, Li H, Thomas-Crusells J, Lahtinen H, Viitanen T, Nanobashvili A, Kokaia Z, Airaksinen MS, Voipio J, Kaila K, Saarma M (2002) BDNF-induced TrkB activation down-regulates the K^+-Cl^- cotransporter KCC2 and impairs neuronal Cl^- extrusion. J Cell Biol 159:747–752

Rivera C, Voipio J, Thomas-Crusells J, Li H, Emri Z, Sipila S, Payne JA, Minichiello L, Saarma M, Kaila K (2004) Mechanism of activity-dependent downregulation of the neuron-specific K-Cl cotransporter KCC2. J Neurosci 24:4683–4691

Samad TA, Moore KA, Sapirstein A, Billet S, Allchorne A, Poole S, Bonventre JV, Woolf CJ (2001) Interleukin-1beta-mediated induction of Cox-2 in the CNS contributes to inflammatory pain hypersensitivity. Nature 410:471–475

Scholz J, Broom DC, Youn DH, Mills CD, Kohno T, Suter MR, Moore KA, Decosterd I, Coggeshall RE, Woolf CJ (2005) Blocking caspase activity prevents transsynaptic neuronal apoptosis and the loss of inhibition in lamina II of the dorsal horn after peripheral nerve injury. J Neurosci 25:7317–7323

Sherman SE, Loomis CW (1994) Morphine insensitive allodynia is produced by intrathecal strychnine in the lightly anesthetized rat. Pain 56:17–29

Sherman SE, Luo L, Dostrovsky JO (1997) Spinal strychnine alters response properties of nociceptive-specific neurons in rat medial thalamus. J Neurophysiol 78:628–637

Sivilotti L, Woolf CJ (1994) The contribution of $GABA_A$ and glycine receptors to central sensitization: disinhibition and touch-evoked allodynia in the spinal cord. J Neurophysiol 72:169–179

Somers DL, Clemente FR (2002) Dorsal horn synaptosomal content of aspartate, glutamate, glycine and GABA are differentially altered following chronic constriction injury to the rat sciatic nerve. Neurosci Lett 323:171–174

Sorkin LS, Puig S (1996) Neuronal model of tactile allodynia produced by spinal strychnine: effects of excitatory amino acid receptor antagonists and a μ-opiate receptor agonist. Pain 68:283–292

Torsney C, Macdermott AB (2006) Disinhibition opens the gate to pathological pain signaling in superficial neurokinin 1 receptor-expressing neurons in rat spinal cord. J Neurosci 26:1833–1843

Trang T, Beggs S, Salter MW (2006) Purinoceptors in microglia and neuropathic pain. Pflugers Arch 452:645–652

Trang T, Beggs S, Wan X, Salter MW (2009) P2X4-receptor mediated synthesis and release of brain-derived neurotrophic factor in microglia is dependent on calcium and p38-mitogen-activated protein kinase activation. J Neurosci 29:3518–3528

Tsuda M, Shigemoto-Mogami Y, Koizumi S, Mizokoshi A, Kohsaka S, Salter MW, Inoue K (2003) P2X4 receptors induced in spinal microglia gate tactile allodynia after nerve injury. Nature 424:778–783

Ulmann L, Hatcher JP, Hughes JP, Chaumont S, Green PJ, Conquet F, Buell GN, Reeve AJ, Chessell IP, Rassendren F (2008) Up-regulation of P2X4 receptors in spinal microglia after peripheral nerve injury mediates BDNF release and neuropathic pain. J Neurosci 28:11263–11268

Vergnano AM, Salio C, Merighi A (2004) NK1 receptor activation leads to enhancement of inhibitory neurotransmission in spinal substantia gelatinosa neurons of mouse. Pain 112:37–47

Walker JM, Hohmann AG (2005) Cannabinoid mechanisms of pain suppression. Handb Exp Pharmacol 509–554

Wang XL, Zhang HM, Li DP, Chen SR, Pan HL (2006) Dynamic regulation of glycinergic input to spinal dorsal horn neurones by muscarinic receptor subtypes in rats. J Physiol 571:403–413

Watanabe M, Wake H, Moorhouse AJ, Nabekura J (2009) Clustering of neuronal K^+-Cl^- cotransporters in lipid rafts by tyrosine phosphorylation. J Biol Chem 284(41):27980–27988

Yaksh TL (1989) Behavioral and autonomic correlates of the tactile evoked allodynia produced by spinal glycine inhibition: effects of modulatory receptor systems and excitatory amino acid antagonists. Pain 37:111–123

Zeilhofer HU, Brune K (2006) Analgesic strategies beyond the inhibition of cyclooxygenases. Trends Pharmacol Sci 27:467–474

Zeilhofer HU, Mohler H, Di LA (2009) GABAergic analgesia: new insights from mutant mice and subtype-selective agonists. Trends Pharmacol Sci 30:397–402

Chapter 8
Injury Induced Epileptogenesis: Contribution of Active Inhibition, Disfacilitation and Deafferentation to Seizure Induction in Thalamocortical System

Igor Timofeev

Abstract Neocortical seizures are the seizures in which neocortex is the leading structure. They are characterized by spike-wave (SW) or spike-wave/polyspike-wave (SW/PSW) complexes of 1.5–3 Hz, intermingled with episodes of fast runs at ~10–20 Hz. These seizures often develop during slow-wave sleep, transition from wake to slow-wave sleep or transition from slow-wave sleep to waking state. Intracellular studies on both anesthetized and non-anesthetized cats have shown that hyperpolarizing phase of the slow oscillation, a distinct feature of slow-wave sleep, is associated with disfacilitation, a temporal absence of synaptic activity in all cortical neurons. Periods of disfacilitation temporally increase network excitability. The hyperpolarizing components of SW-PSW complexes are mediated mainly by leak current (state similar to discfacilitation), Ca^{2+}- and Na^{+}-activated K^{+} currents. It is proposed that prolonged periods of disfacilitation up-regulate neuronal excitability that contributes to the seizure generation. Once seizure has started, fast-spiking inhibitory interneurons fire multiple action potentials during paroxysmal depolarizing shifts (EEG spike components of SW/PSW complexes). During seizure a set of cellular processes induces a shift of reversal potential of GABA toward depolarization. Intense firing of GABAergic neurons and depolarizing GABA responses largely contribute to the generation of paroxysmal EEG spikes. Inhibition does not play a role in other components of neocortical seizures. Neocortical trauma, in particular penetrating wounds, produces partial deafferentation of a subset of neurons that decreases excitability of network. Cortical neurons display occasional periods of disfacilitation in deafferented cortex during all states of vigilance. As in the case of slow-wave sleep, periods of disfacilitation up-regulate neuronal excitability. Synaptic volleys originating from preserved axons impinge hyperexcitable neurons of deafferented cortex that trigger seizures.

I. Timofeev (✉)
Department of Psychiatry and Neuroscience, Laval University, The Centre de recherche Université Laval Robert-Giffard (CRULRG), 2601, chemin de la Canardière, Québec G1J 2G3, Canada
e-mail: Igor.Timofeev@phs.ulaval.ca

M.A. Woodin and A. Maffei (eds.), *Inhibitory Synaptic Plasticity*,
DOI 10.1007/978-1-4419-6978-1_8, © Springer Science+Business Media, LLC 2011

I propose that any physiological or pathological condition that leads to repeated or prolonged periods of neuronal silence will increase neuronal hyperexcitability that favours development of seizures.

Introduction

All brain functions are executed during three distinct states of vigilance: wake, slow-wave sleep (SWS) and rapid-eye movement (REM) sleep. Each of these states is characterized by particular patterns of oscillatory activities (Bazhenov and Timofeev 2006) that are generated by thalamocortical system and may be recorded as field potentials from cortical surface or using intracellular recordings. A distinct pathological condition within thalamocortical network is paroxysmal or epileptiform activity that can lead to development of seizures (Niedermeyer 2005). During seizures, both cortical and thalamic neurons display paroxysmal membrane potential excursions. A vast majority of thalamocortical neurons do not fire action potentials during neocortical seizures and thus do not contribute actively to the generation of neocortical seizures (reviewed in (Timofeev 2010)). Therefore the main focus of this chapter will be devoted to neocortical processes accompanying epileptogenesis and seizures.

Cortical Neuronal Activities During States of Vigilance

Membrane potential of neurons within thalamocortical loop reveals distinct patterns of activity during waking state, SWS and REM sleep and it is highly correlated with EEG signal. EEG pattern is activated (high frequency, low amplitude waves) during waking state and REM sleep (Moruzzi and Magoun 1949), while SWS is characterized by generation of large amplitude slow waves of EEG (Blake and Gerard 1937). Experiments on anesthetized animals show that during EEG depth-positivity, cortical neurons remain in hyperpolarized, silent state, while during EEG depth-negativity cortical neurons move to active states, reveal barrages of synaptic events and fire action potentials (Contreras and Steriade 1995). The same patterns are also found during natural sleep (Steriade et al. 2001; Timofeev et al. 2001). Intracellular studies on both anesthetized and non-anesthetized cats have shown that hyperpolarizing phase of the slow oscillation is associated with disfacilitation, a temporal absence of synaptic activity in all cortical neurons (Contreras et al. 1996; Timofeev et al. 1996, 2001). During waking state, REM sleep and depolarizing phases of SWS both excitatory and inhibitory cortical neurons are relatively depolarized and usually fire action potentials. A certain balance of excitation and inhibition is needed to maintain active network states (Shu et al. 2003; Rudolph et al. 2007; Haider and McCormick 2009). However, fast spiking inhibitory interneurons fire action potentials at higher rates as compare to pyramidal cells

(Steriade et al. 2001). This imposes important inhibitory influence on their targets; during quiet wakefulness, almost a half of regular-spiking neurons (25% of total number of neurons) are relatively hyperpolarized, reveal multiple trains of IPSPs and do not fire action potentials (Rudolph et al. 2007). Overall, during wakefulness, REM sleep and active states of SWS, the membrane-potential activity stems from a combination of excitatory and inhibitory synaptic conductances, with dominant inhibition in most of the cases (Rudolph et al. 2007).

Neocortical Seizures and Active Inhibition

Neocortical seizures are the seizures in which neocortex is the leading structure. They are characterized by spike-wave (SW) or spike-wave/polyspike-wave (SW/PSW) complexes of 1.5–3 Hz, intermingled with episodes of fast runs at ~7–20 Hz (Steriade et al. 1998; Timofeev et al. 1998; Boucetta et al. 2008). Spontaneously occurring SW complexes at 1–2.5 Hz and fast runs at 7–20 Hz develop without discontinuity from slow (mainly 0.5–0.9 Hz) cortically generated oscillations (Steriade and Contreras 1995). At the focus, the onset of neocortical seizures is accompanied with generation of high frequency oscillations (>100 Hz) termed ripples, which are not over expressed in other cortical regions at this time (Grenier et al. 2003a). During seizure, the ripples can be recorded at multiple cortical locations. Because of the high likelihood of seizure development from slow oscillations, neocortical seizures very often develop during SWS, transition from wake to SWS or transition from SWS to waking state. Therefore most of neocortical seizures belong to a class of nocturnal seizures. SW/PSW complexes are associated with clonic components of seizures and runs of fast EEG spikes that last longer than 2–3 s are accompanied with tonic components of seizures (Niedermeyer 2005). Similar to SWS, both excitatory and inhibitory cortical neurons are depolarized and fire action potentials during EEG spikes and are hyperpolarized and silent during EEG waves. The hyperpolarizing components of SW-PSW complexes are mediated mainly by leak current (state similar to discfacilitation), Ca^{2+}- and Na^{+}-activated K^{+} currents (Timofeev et al. 2004). In about 20% of neurons, a hyperpolarization-activated depolarizing current (I_H) is also active and its activity leads network to the generation of next paroxysmal cycles. Altogether, this indicates that hyperpolarizing or "EEG-wave" components of seizures are not generated by activity of inhibitory neurons. By contrast, active inhibition is implicated in the generation of depolarizing "EEG-spike" components of SW/PSW seizures. This is demonstrated by the facts that during EEG-spikes, expressed as paroxysmal depolarizing shifts at intracellular level: (1) the fast-spiking inhibitory interneurons fire high-frequency (up to 800 Hz) spike trains, while regular-spiking neurons fire usually one or very few action potentials, (2) synaptic events easily reversed at depolarizing potentials, (3) the amplitude of paroxysmal depolarizing shifts is greatly enhanced by intracellular recordings performed with chloride containing pipettes and finally (4) antagonists of GABAergic signalling abolished synchronous interictal discharges

(Cohen et al. 2002; Timofeev et al. 2002; Timofeev and Steriade 2004). Inhibitory interneurons fire very few spikes, if any, during paroxysmal fast runs (Timofeev et al. 1998) and therefore do not play a leading role in the generation of this type of paroxysmal activities. Despite high frequency firing of interneurons during paroxysmal depolarizing shifts, during which paroxysmal ripple activity can be recorded from local field potential electrodes, the individual ripple waves and the firing of inhibitory interneurons are not synchronized (Grenier et al. 2003a). This stands in contrast with perfect synchrony of action potentials of fast-spiking inhibitory cells with individual ripple waves during normal slow oscillation (Grenier et al. 2001).

Synchronous cortical drives during SW/PSW complexes induce burst firing of inhibitory reticular thalamic neurons that inhibit thalamocortical cells for the duration of SW/PSW complexes (Steriade and Contreras 1995; Pinault et al. 1998; Timofeev et al. 1998; Polack and Charpier 2006). Thalamocortical neurons do not fire action potentials and thus do not contribute actively to neocortical seizure generation.

Overall, fast-spiking inhibitory interneurons contribute to the generation of paroxysmal depolarizing shifts of EEG spike components of SW/PSW complexes, but do not play a major role in the generation of other components of neocortical seizures.

Epileptogenesis

Epileptogenesis is a hidden continuous process occurring between the initial cortical insult and the explicit onset of late epilepsy (Fig. 8.1). The trademark of epilepsy is the expression of unprovoked seizures, characterized by paroxysmal neuronal activities and high synchronization. Trauma and brain infection are the primary source of acquired epilepsy, they can cause epilepsy at any age, and may account for a higher incidence of epilepsy in developing countries. Traumatic brain

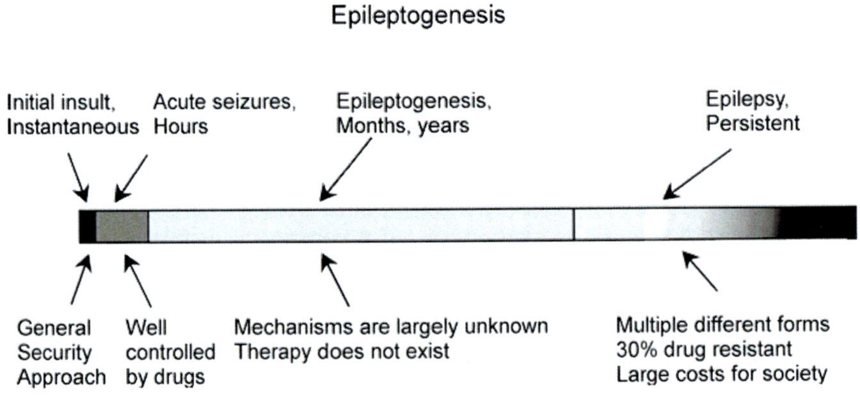

Fig. 8.1 Time course of development of epilepsy from brain trauma

injury is a major risk factor for subsequent development of epilepsy (Feeney and Walker 1979; Temkin et al. 1995; Annegers et al. 1998). Acute cerebral cortical trauma leads to paroxysmal activities. This is a normal response of brain to the injury. Within 24 h up to 80% of patients with penetrating wounds display clinical seizures that stop within a 48 h period (Kollevold 1976; Dinner 1993). In human, epileptogenesis develops over several months or years (Fig. 8.1). Vietnam and Croatia postwar epidemiological studies report that 10–15 years after the trauma about 50% of patients with penetrating cranial wounds develop epilepsy characterized by spontaneous recurring seizures (Salazar et al. 1985; Marcikic et al. 1998). Trauma induced epilepsy is poorly controlled by currently available medication. Early administration of anticonvulsant medication decreases the percentage of early posttraumatic seizures but not that of chronic epilepsy (Temkin et al. 1990; Temkin et al. 1999; Chang and Lowenstein 2003). The mechanisms of epileptogenesis are not understood. The current treatment is designed to control epilepsy or seizures, but not epileptogenesis.

Current Experimental Models of Trauma-Induced Epileptogenesis: (Cortical Undercuts and Slabs)

Two major factors with different impacts accompany trauma-induced epilepsy: mechanical and hemorrhage components. In some forms of trauma, a resulting hematoma can play a key role in posttraumatic epilepsy (D'Ambrosio et al. 2004; Messori et al. 2005). This is confirmed by the fact that injection of iron salts produces electrographic and behavioral seizures (Willmore et al. 1978). However, penetrating cortical wounds usually do not induce large hematoma, but still lead to epilepsy. Below, I explore possible mechanisms of trauma-induced epilepsy with little if any hematoma. I propose that in this type of epileptogenesis, the deafferentation plays a leading role in triggering a number of events that up-regulate network excitability and trigger seizures.

Early studies in both human and animals demonstrated that isolated cortical islands or cortical slabs develop abnormal hyperexcitability (Echlin et al. 1952; Sharpless and Halpern 1962; Echlin and Battista 1963; Sharpless 1969). In vivo recordings from acute cortical slabs demonstrated the presence of long-lasting (tens of sec) periods of silence, interrupted by brief (about 1 s) periods of activity (Timofeev et al. 2000b). These active periods could be easily elicited by low intensity electrical stimulation (Timofeev et al. 2000b). In vitro recordings from chronically isolated slabs demonstrated the presence of similar events (Prince and Tseng 1993), which were called epileptiform discharges. Chronically isolated cortical slabs in non-anesthetized cats showed the presence of slow rhythmic activity independent of brain states (Houweling et al. 2005). Rhythmic electrical stimulation of slab was able to elicit electrographic seizure composed of fast runs and SW/PSW discharges (Timofeev et al. 1998). This indicates that acute isolated cortical slabs can generate seizures, but generate them only when external input is present.

Partially isolated neocortex (Fig. 8.2) is a well-established model of trauma-induced epilepsy (Hoffman et al. 1994; Prince et al. 1997; Jacobs et al. 2000; Li and Prince 2002; Topolnik et al. 2003a, b; Li et al. 2005; Nita et al. 2006, 2007). In these preparations thalamocortical, callosal and some other connections are severed, but some horizontal, cortico-cortical connections remain intact that provide sufficient

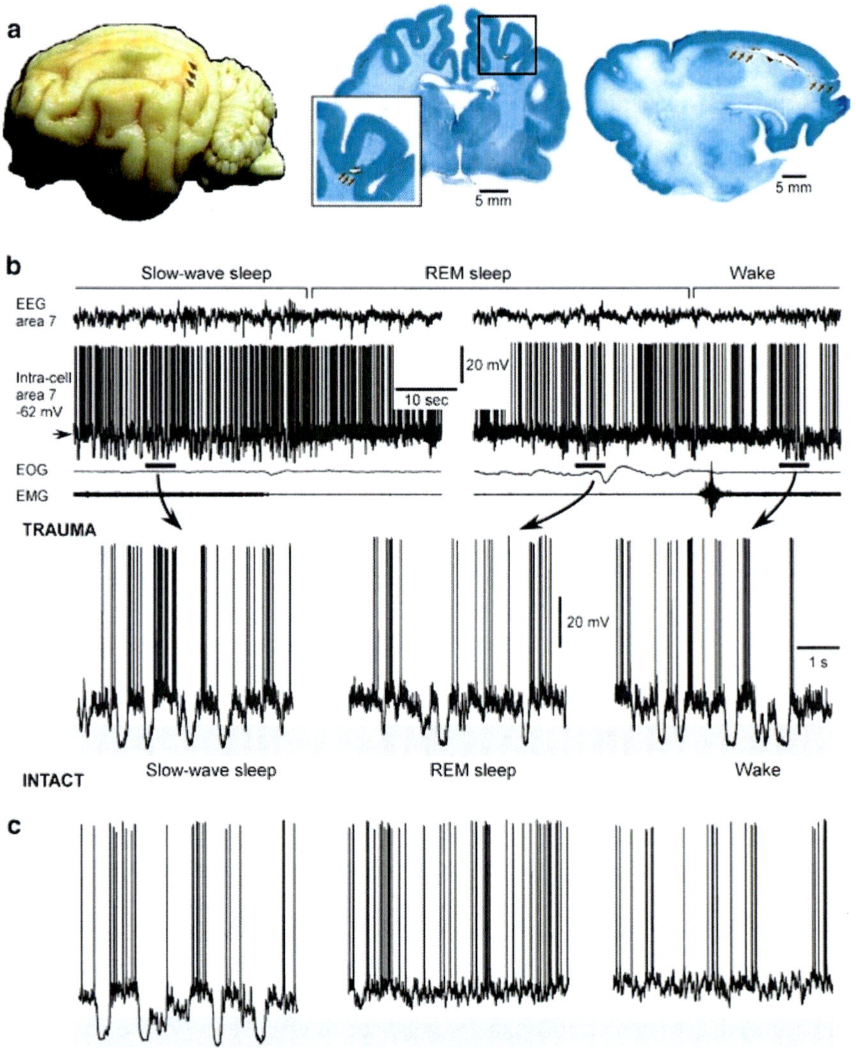

Fig. 8.2 Long-lasting silent periods occur in all states of vigilance in the undercut cortex. (**a**) Cat brain depicting the localization of the undercut (*arrows*), global view – *left*, frontal section – *middle* and sagital section – *right*. (**b**) Intracellular and field potential recordings during different states of vigilance. Epochs indicated by horizontal bars are expanded below. Note the presence of large amplitude hyperpolarizing potentials, indicated by shadowed area, during REM sleep and waking state. (**c**) Intracellular neuronal recording in the intact cortex. (Form Timofeev et al. 2010)

external inputs to deafferented (hyper-excitable, see below) part of cortex enabling generation of seizures. Our previous in vivo studies demonstrated that paroxysmal activities induced by cortical undercut share major features of human penetrating wound induced epilepsy. There features are: (a) Immediately after undercut the activity in traumatized cortex is depressed (Topolnik et al. 2003a), (b) paroxysmal discharges appear in 2–3 h after undercut in 80% of experimental animals and usually terminates 5–8 h later (Topolnik et al. 2003a). By means of local field potentials, extracellular unit, and intracellular activities (not possible in human brain), we demonstrated additionally that acute seizures started at the border between intact and undercut cortex and propagated toward more damaged parts of the cortical tissue (Topolnik et al. 2003a). We have shown that (1) increase in the synchrony toward the end of seizures induces larger neuronal depolarization, activates Ca^{2+}-activated K^{+} currents and terminate seizures (Timofeev et al. 2004; Timofeev and Steriade 2004; Timofeev and Bazhenov 2005), (2) at the border of intact and undercut cortex the intrinsic excitability of neurons and the network synaptic excitability is increased (Avramescu and Timofeev 2008; Topolnik et al. 2003b), (3) in the same region of cortex, there is a positive shift in the IPSP reversal potential (Topolnik et al. 2003b). In chronic experiments, using multisite field potential recordings in non-anesthetized cats, we demonstrated that electrographic seizures reappear at the border of intact and undercut cortex (Nita et al. 2007). With time, paroxysmal activities involve neighboring cortical regions and at about 4–6 weeks after the undercut, paroxysmal discharges are recorded in all electrodes. Like in human, these paroxysmal discharges disappeared during REM sleep (Nita et al. 2007). Some animals start to display behavioral paroxysmal events 30 days after surgery. In our video recordings, we observed (1) absence type seizures, and (2) repetitive paroxysmal jerks. Recordings from anesthetized animals with undercut display progressive increase in field potential amplitudes in the undercut cortex and propagation of activity from the border of intact cortex toward undercut area (Topolnik et al. 2003a; Nita et al. 2006). The most unexpected finding in these experiments came from intracellular recordings from non-anesthetized animals. Previously, we have shown that during SWS in normal animals (no undercut) neocortical neurons reveal active (usually depolarizing or UP) and silent (always hyperpolarizing or DOWN) states. During both waking state and REM sleep, cortical neurons remain in active state (Timofeev et al. 2000a, 2001; Steriade et al. 2001; Mukovski et al. 2007; Rudolph et al. 2007). *Intracellular recordings from partially deafferented cortex during both waking state and even REM sleep revealed the presence of silent states that are similar to silent states (periods of disfacilitation) observed during SWS* (Fig. 8.2) (Nita et al. 2007; Timofeev et al. 2010). We propose that removal of a part of excitatory inputs produced by the undercut of white matter temporally decreases network excitation; in these conditions cortical network cannot maintain persistent active state that leads to the occurrence of silent states.

A critical number of neurons is essential in maintaining persistent active states. As it was stated previously, during SWS cortical network oscillates between active and silent states with a frequency around the 1 Hz, while during brain activated states, cortical neurons are permanently active. In acute neocortical slabs of cats with an estimated number of neurons of ten million (Timofeev et al. 2000b), the active states were generated only occasionally and their duration did not exceed several seconds. Intracellular recordings from striatal medium spiny neurons of rats (Mahon et al. 2006) demonstrated the same pattern of activity in these neurons as in cortical neurons of cats and indirectly suggest that cortical network of rats and cats behave in a similar manner during waking and sleeping states. However, in mice, in which the cortical network is composed of a significantly smaller number of neurons, patch-clamp recordings during passive waking state demonstrated unstable membrane potential, oscillating between active and silent states and stable levels of membrane potential were present only during active wakefulness, when additional excitatory synaptic drive reached neocortex (Crochet and Petersen 2006; Poulet and Petersen 2008; Gentet et al. 2010).

After several weeks of cortical permanent partial deafferentaion, achieved via undercutting underlying white matter, the following morphological and physiological changes take place: (a) reduction of cortical thickness, accompanied with delamination and particular reduction of deep layer neurons (Avramescu et al. 2009), (b) a preferential loss of inhibitory GABAergic neurons as opposed to glutamatergic excitatory neurons (Avramescu et al. 2009), (c) despite that fact the probability of connections between neighboring neurons increases (Avramescu and Timofeev 2008), suggesting intense sprouting of cortical axons (Salin et al. 1995), (d) an increase in input resistance and intrinsic neuronal excitability of regular-spiking neurons (Avramescu and Timofeev 2008), likely due to a total reduction in the number of synapses and/or synaptic activity in the deafferented cortex, and (e) increased spontaneous burst firing (Nita et al. 2006). In vitro neuronal recording performed from chronically undercut cortex in other laboratories reveal similarly increased intrinsic and synaptic excitability (Prince and Tseng 1993; Hoffman et al. 1994; Prince et al. 1997; Li and Prince 2002; Jin et al. 2006). They also suggested an impaired Cl^- metabolism in neurons from undercut cortex (Jin et al. 2005).

Homeostatic Plasticity and Trauma Induced Epileptogenesis

When cellular cultures are kept for two days in sodium channel blockers, the amplitude of spontaneous spike-independent synaptic events (minis) is increased as compare to control cultures and vice versa, keeping cultures in medium containing GABA receptor blockers (a measure that increases network activity) decreases the amplitude of minis (Turrigiano et al. 1998). Intrinsic neuronal excitability is also regulated by activity. After chronic activity blockade, Na^+ currents increase and K^+ currents decrease in size, resulting in an enhanced responsiveness of pyramidal cells to current injections (Desai et al. 1999). These observations suggest the existence of a fundamental mechanism, termed "homeostatic plasticity" (Turrigiano 1999), that controls the levels of neuronal activity, such as the firing rates (Turrigiano et al. 1998; Murthy et al. 2001). These observations led us to formulate a hypothesis that homeostatic plasticity induced by trauma triggers epileptogenesis

(Houweling et al. 2005). Deafferentation leads to a reduction of synaptic inputs to affected cortex. This condition by itself may be a sufficient factor for the up-regulation of synaptic efficacy. An additional factor up-regulating synaptic efficacy is the presence of prolonged hyperpolarizing (silent) states, which lasts longer in deafferented cortex than in intact cortical regions and/or in control animals (Topolnik et al. 2003a; Avramescu and Timofeev 2008; Timofeev et al. 2010). The presence of these two factors induces a compensatory boost in neuronal excitability that leads to seizure generation.

Some Other Related Models of Interest

(1) Several weeks or months after pyramidal tract lesion the responsiveness of corticospinal neurons was enhanced and this was accompanied with axonal sprouting of layer 5 neurons (Tseng and Prince 1996). (2) Similarly to undercut, chronic GABA infusion suppresses activity in the region of infusion and upon withdrawal from chronic GABA, paroxysmal activities occur (Brailowsky et al. 1988, 1989; Silva-Barrat et al. 1992). (3) Neocortical kindling is most efficient when electrical stimulation is applied during transition from SWS to waking state, suggesting that the presence of silent network states may be favorable in the development of epilepsy (Nita et al. 2008a, b). Seizures induced by neocortical kindling are followed by outlasting activities that appear as alterations of active and silent network states (see Fig. 8.3, upper right panel) and similarly to trauma induced epileptogenesis are associated with increased neuronal bursting (Nita et al. 2008a). Overall, this suggests that reduced activity in some cortical territories for days, weeks, or months and/or increase in the duration or number of silent states favor the occurrence of seizures.

Epileptogenesis and the Onset of Seizures

The neocortex is an important component in many forms of paroxysmal activity and it is actively involved in the generation of paroxysmal discharges (Steriade and Contreras 1995; Pinault et al. 1998; Steriade et al. 1998; Steriade and Contreras 1998; Timofeev et al. 1998, 2002a, b, 2004; Crunelli and Leresche 2002; Meeren et al. 2002; Timofeev and Steriade 2004). However, the mechanisms underlying neocortical epileptogenesis remain largely unknown.

Seizures induced by deafferentation (cortical undercut (Topolnik et al. 2003a, b; Nita et al. 2006)) and nocturne seizures (seizures evolving from sleep oscillations (Neckelmann et al. 1998; Steriade et al. 1998; Steriade and Contreras 1998; Timofeev et al. 1998)) share some similarities. The state of the SWS is characterized by the presence of long-lasting periods of disfacilitation associated with neuronal hyperpolarization (Timofeev et al. 2001). During periods of disfacilitation synapses are likely released from a steady-state depression (Galarreta and Hestrin 1998)

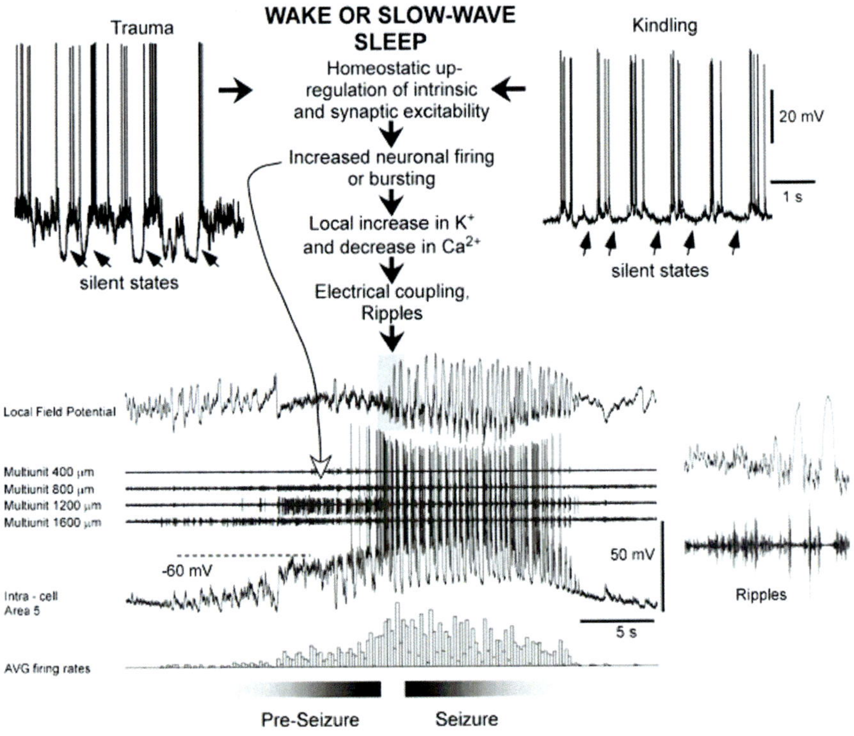

Fig. 8.3 Epileptogenesis leading to onset of seizure activity. *Upper panels* – both trauma and kindling induce long-lasting alterations of active and silent states that occur during slow-wave sleep and waking states. Silent states are indicated by *arrows*. Prolonged silent states contribute to the induction of homeostatic up-regulation of intrinsic and synaptic excitability that increases neuronal firing or bursting. Increase neuronal activity provokes local increase in extracellular K^+ and decrease in extracellular Ca^{2+} concentrations that favor electrical coupling. If some threshold over passed, pathological ripples are generated that lead to development of seizure. *Lower panel* – progressively increased neuronal firing prior to seizure and decreased neuronal firing toward the end of the seizure. The seizure is composed of typical spike-and-wave complexes. Fragment indicated by *grey box* is expanded in *right panel*. In this panel the lower trace is the same local field potential trace filtered between 80 and 200 Hz to show ripples. (*Upper Left panel* – from Fig. 8.2; *Upper Right panel* – from Nita et al. 2008a; *Lower panel* – from Bazhenov et al. 2004, with modifications)

and the synaptic transmission is strengthen that could contribute to the onset of seizures. The penetrating wounds or acute experimental deafferentation has been described as strong epileptogenic factors (Kollevold 1976; Dinner 1993; Prince et al. 1997; Topolnik et al. 2003a, b; Jacobs and Prince 2005; Jin et al. 2005). In such conditions, some of the axons impinging onto postsynaptic neurons are damaged and not functioning properly, which creates a partial deafferentation that may increase the sensitivity of cortical neurons in those foci and in surrounding areas (Abbott et al. 1997; Desai et al. 1999a, b). The synaptic inputs from cortical regions exhibiting moderate or high levels of activity would then lead to an increased responsiveness in those cortical areas where the sensitivity is increased. Thus, both factors, the sleep-related disfacilitation and the traumatic deafferentation, may increase the probability of seizures via the same mechanism of an increased effectiveness of synaptic transmission (Crochet et al. 2005; Li et al. 2005). The increased effectiveness of synaptic transmission depends on higher levels of extracellular Ca^{2+} concentration during silent periods of network activities as opposed to active network states (Massimini and Amzica 2001; Crochet et al. 2005), and on synaptic facilitation that follows periods of neuronal silence (Galarreta and Hestrin 1998). Decreased level of extracellular Ca^{2+} increases the intrinsic excitability of cortical neurons and convert some of them to burst firing (Boucetta et al. 2003). In parallel, decreased extracellular Ca^{2+} opens hemi-channels forming gap junctions, and thus increases electrical coupling (Thimm et al. 2005). The synaptic excitability is also enhanced by a trauma-related increase in glutamate levels (Sakowitz et al. 2002). The neuronal firing increases at the onset of seizures (Bazhenov et al. 2004), which generates ripple activity that is highest at the onset in the focus of seizures (Fig. 8.3) (Grenier et al. 2003a, b; Timofeev and Steriade 2004). Intense neuronal firing increases extracellular K^+ concentration that further boost neuronal firing. In trauma-related acute seizures, however, factors other than synaptic could be considered as promoting seizures. These factors primarily depend on increase in extracellular levels of K^+ (Moody et al. 1974) that lead to enhanced intrinsic excitability of neurons (Traynelis and Dingledine 1988; McNamara 1994; Timofeev et al. 2002; Topolnik et al. 2003b). Increased neuronal firing, further increases extracellular levels of K^+ (Traynelis and Dingledine 1988), and decreases extracellular levels of Ca^{2+} (Pumain et al. 1983) that leads to strengthening of field potential coupling between neurons (Grenier et al. 2003b), which together leads to the generation of paroxysmal ripple activity and as consequence to seizure generation (Fig. 8.3).

Perspectives

Understanding the cellular and molecular mechanisms of hyperexcitability induced by deafferentation will lead to the invention of precise pharmacological tools that would prevent development of epileptogenesis. However, we are only at the beginning of this process. From the other hand, it is obvious that increased silence is a key factor of trauma induced epileptogenesis. In the short term, a decrease or abolishment of neuronal silence induced by stimulation will not restore trauma-dependent loss of function, but might prevent epiletogenesis. The use of random electrical stimulation of deafferented area may be a promising tool, but it may kindle the stimulated area, worsening epileptogenic conditions. An alternative approach would be to use viral insertion of channelrhodopsine-2 (Zhang et al. 2006) combined with light stimulation or to use any local pharmacological stimulation that diminishes hyperpolarizing K^+ currents (e.g. cholinergic agonists). The use of these locally applied tools may be easy in the case of penetrating wounds, when the damaged area can be accessed from the side of the wound. The stimulation of damaged area must start as soon as acute seizures are terminated and last at least for months or years. This measure will prevent epileptogenesis and these patients will not develop epilepsy at later stages.

Acknowledgments My research is supported by Canadian Institutes of Health Research, National Institutes of Health, Fonds de la Recherche en Santé du Québec and Natural Science and Engineering Research Council of Canada.

References

Abbott LF, Varela JA, Sen K, Nelson SB (1997) Synaptic depression and cortical gain control. Science 275:220–224.

Annegers JF, Hauser WA, Coan SP, Rocca WA (1998) A population-based study of seizures after traumatic brain injuries. N Engl J Med 338:20–24.

Avramescu S, Timofeev I (2008) Synaptic strength modulation following cortical trauma: a role in epileptogenesis. J Neurosci 28:6760–6772.

Avramescu S, Nita D, Timofeev I (2009) Neocortical post-traumatic epileptogenesis is associated with the loss of GABAergic neurons. J Neurotrauma 26:799–812.

Bazhenov M, Timofeev I (2006) Thalamocortical oscillations. In: http://www.scholarpedia.org/article/Thalamocortical_Oscillations.

Bazhenov M, Timofeev I, Steriade M, Sejnowski TJ (2004) Potassium model for slow (2–3 Hz) in vivo neocortical paroxysmal oscillations. J Neurophysiol 92:1116–1132.

Blake H, Gerard RW (1937) Brain potentials during sleep. Am J Physiol 119:692–703.

Boucetta S, Crochet S, Timofeev I (2003) Effects of extracellular Ca 2+ concentration on firing properties of cortical neurons. In: SFN annual meeting, p Program No. 791.796. New Orleans.

Boucetta S, Chauvette S, Bazhenov M, Timofeev I (2008) Focal generation of paroxysmal fast runs during electrographic seizures. Epilepsia 49:1925–1940.

Brailowsky S, Kunimoto M, Menini C, Silva-Barrat C, Riche D, Naquet R (1988) The GABA-withdrawal syndrome: a new model of focal epileptogenesis. Brain Res 442:175–179.

Brailowsky S, Silva-Barrat C, Menini C, Riche D, Naquet R (1989) Effects of localized, chronic GABA infusions into different cortical areas of the photosensitive baboon, Papio papio. Electroencephalogr Clin Neurophysiol 72:147–156.

Chang BS, Lowenstein DH (2003) Practice parameter: antiepileptic drug prophylaxis in severe traumatic brain injury: report of the quality standards subcommittee of the American academy of neurology. Neurology 60:10–16.

Cohen I, Navarro V, Clemenceau S, Baulac M, Miles R (2002) On the origin of interictal activity in human temporal lobe epilepsy in vitro. Science 298:1418–1421.

Contreras D, Steriade M (1995) Cellular basis of EEG slow rhythms: a study of dynamic corticothalamic relationships. J Neurosci 15:604–622.

Contreras D, Timofeev I, Steriade M (1996) Mechanisms of long-lasting hyperpolarizations underlying slow sleep oscillations in cat corticothalamic networks. J Physiol 494:251–264.

Crochet S, Petersen CC (2006) Correlating whisker behavior with membrane potential in barrel cortex of awake mice. Nat Neurosci 9:608–610.

Crochet S, Chauvette S, Boucetta S, Timofeev I (2005) Modulation of synaptic transmission in neocortex by network activities. Eur J Neurosci 21:1030–1044.

Crunelli V, Leresche N (2002) Childhood absence epilepsy: genes, channels, neurons and networks. Nat Rev Neurosci 3:371–382.

D'Ambrosio R, Fairbanks JP, Fender JS, Born DE, Doyle DL, Miller JW (2004) Post-traumatic epilepsy following fluid percussion injury in the rat. Brain 127:304–314.

Desai NS, Nelson SB, Turrigiano GG (1999) Activity-dependent regulation of excitability in rat visual cortical neurons. Neurocomputing 26–27:101–106.
Desai NS, Rutherford LC, Turrigiano GG (1999) Plasticity in the intrinsic excitability of cortical pyramidal neurons. Nature Neuroscience 2:515–520.
Dinner D (1993) Posttraumatic epilepsy. In: The Treatment of Epilepsy: Principles (E. W, ed), pp 654–658. Philadelphia: Lea & Fibinger.
Echlin FA, Battista A (1963) Epileptiform seizures from chronic isolated cortex. Arch Neurol 168:154–170.
Echlin FA, Arnet V, Zoll J (1952) Paroxysmal high voltage discharges from isolated and partially isolated human and animal cerebral cortex. Electroencephalogr Clin Neurophysiol 4:147–164.
Feeney DM, Walker AE (1979) The prediction of posttraumatic epilepsy. A mathematical approach. Arch Neurol 36:8–12.
Galarreta M, Hestrin S (1998) Frequency-dependent synaptic depression and the balance of excitation and inhibition in the neocortex. Nat Neurosci 1:587–594.
Gentet LJ, Avermann M, Matyas F, Staiger JF, Petersen CCH (2010) Membrane potential dynamics of GABAergic neurons in the barrel cortex of behaving mice. Neuron 65:422–435.
Grenier F, Timofeev I, Steriade M (2001) Focal synchronization of ripples (80–200 Hz) in neocortex and their neuronal correlates. J Neurophysiol 86:1884–1898.
Grenier F, Timofeev I, Steriade M (2003a) Neocortical very fast oscillations (ripples, 80–200 Hz) during seizures: intracellular correlates. J Neurophysiol 89:841–852.
Grenier F, Timofeev I, Crochet S, Steriade M (2003b) Spontaneous field potentials influence the activity of neocortical neurons during paroxysmal activities in vivo. Neuroscience 119:277–291.
Haider B, McCormick DA (2009) Rapid neocortical dynamics: cellular and network mechanisms. Neuron 62:171–189.
Hoffman SN, Salin PA, Prince DA (1994) Chronic neocortical epileptogenesis in vitro. J Neurophysiol 71:1762–1773.
Houweling AR, Bazhenov M, Timofeev I, Steriade M, Sejnowski TJ (2005) Homeostatic synaptic plasticity can explain post-traumatic epileptogenesis in chronically isolated neocortex. Cereb Cortex 15:834–845.
Jacobs KM, Graber KD, Kharazia VN, Parada I, Prince DA (2000) Postlesional epilepsy: the ultimate brain plasticity. Epilepsia 41 Suppl 6:S153-S161.
Jacobs KM, Prince DA (2005) Excitatory and inhibitory postsynaptic currents in a rat model of epileptogenic microgyria. J Neurophysiol 93:687–696.
Jin X, Huguenard JR, Prince DA (2005) Impaired Cl^- extrusion in layer V pyramidal neurons of chronically injured epileptogenic neocortex. J Neurophysiol 93:2117–2126.
Jin X, Prince DA, Huguenard JR (2006) Enhanced excitatory synaptic connectivity in layer V pyramidal neurons of chronically injured epileptogenic neocortex in rats. J Neurosci 26:4891–4900.
Kollevold T (1976) Immediate and early cerebral seizures after head injuries. Part I. J Oslo City Hosp 26:99–114.
Li H, Prince DA (2002) Synaptic activity in chronically injured, epileptogenic sensory-motor neocortex. J Neurophysiol 88:2–12.
Li H, Bandrowski AE, Prince DA (2005) Cortical injury affects short-term plasticity of evoked excitatory synaptic currents. J Neurophysiol 93:146–156.
Mahon S, Vautrelle N, Pezard L, Slaght SJ, Deniau J-M, Chouvet G, Charpier S (2006) Distinct patterns of striatal medium spiny neuron activity during the natural sleep-wake cycle. J Neurosci 26:12587–12595.
Marcikic M, Melada A, Kovacevic R (1998) Management of war penetrating craniocerebral injuries during the war in Croatia. Injury 29:613–618.
Massimini M, Amzica F (2001) Extracellular calcium fluctuations and intracellular potentials in the cortex during the slow sleep oscillation. J Neurophysiol 85:1346–1350.
McNamara JO (1994) Cellular and molecular basis of epilepsy. J Neurosci 14:3413–3425.
Meeren HK, Pijn JP, Van Luijtelaar EL, Coenen AM, Lopes da Silva FH (2002) Cortical focus drives widespread corticothalamic networks during spontaneous absence seizures in rats. J Neurosci 22:1480–1495.

Messori A, Polonara G, Carle F, Gesuita R, Salvolini U (2005) Predicting posttraumatic epilepsy with MRI: prospective longitudinal morphologic study in adults. Epilepsia 46:1472–1481.
Moody WJ, Futamachi KJ, Prince DA (1974) Extracellular potassium activity during epileptogenesis. Exp Neurol 42:248–263.
Moruzzi G, Magoun HW (1949) Brain stem reticular formation and activation of the EEG. Electroencephalogr Clin Neurophysiol 1:455–473.
Mukovski M, Chauvette S, Timofeev I, Volgushev M (2007) Detection of active and silent states in neocortical neurons from the field potential signal during slow-wave sleep. Cereb Cortex 17:400–414.
Murthy VN, Schikorski T, Stevens CF, Zhu Y (2001) Inactivity produces increases in neurotransmitter release and synapse size. Neuron 32:673–682.
Neckelmann D, Amzica F, Steriade M (1998) Spike-wave complexes and fast components of cortically generated seizures. III. Synchronizing mechanisms. J Neurophysiol 80:1480–1494.
Niedermeyer E (2005) Abnormal EEG patterns: epileptic and paroxysmal. In: Electroencephalography: Basic Principles, Clinical Applications, and Related Fields, Fifth Edition Edition (Niedermeyer E, Lopes de Silva F, eds), pp 255–280. Philadelphia: Lippincott Williams & Wilkins.
Nita D, Cisse Y, Timofeev I (2008a) State-dependent slow outlasting activities following neocortical kindling in cats. Exp Neurol 211:456–468.
Nita D, Cissé Y, Timofeev I, Steriade M (2006) Increased propensity to seizures after chronic cortical deafferentation in vivo. J Neurophysiol 95:902–913.
Nita D, Cisse Y, Frohlich F, Timofeev I (2008b) Cortical and thalamic components of neocortical kindling-induced epileptogenesis in behaving cats. Exp Neurol 211:518–528.
Nita DA, Cisse Y, Timofeev I, Steriade M (2007) Waking-sleep modulation of paroxysmal activities induced by partial cortical deafferentation. Cereb Cortex 17:272–283.
Pinault D, Leresche N, Charpier S, Deniau JM, Marescaux C, Vergnes M, Crunelli V (1998) Intracellular recordings in thalamic neurones during spontaneous spike and wave discharges in rats with absence epilepsy. J Physiol 509:449–456.
Polack P-O, Charpier S (2006) Intracellular activity of cortical and thalamic neurones during high-voltage rhythmic spike discharge in Long-Evans rats in vivo. J Physiol 571:461–476.
Poulet JFA, Petersen CCH (2008) Internal brain state regulates membrane potential synchrony in barrel cortex of behaving mice. Nature 454:881–885.
Prince DA, Tseng GF (1993) Epileptogenesis in chronically injured cortex: in vitro studies. J Neurophysiol 69:1276–1291.
Prince DA, Jacobs KM, Salin PA, Hoffman S, Parada I (1997) Chronic focal neocortical epileptogenesis: does disinhibition play a role? Can J Physiol Pharmacol 75:500–507.
Pumain R, Kurcewicz I, Louvel J (1983) Fast extracellular calcium transients: involvement in epileptic processes. Science 222:177–179.
Rudolph M, Pospischil M, Timofeev I, Destexhe A (2007) Inhibition determines membrane potential dynamics and controls action potential generation in awake and sleeping cat cortex. J Neurosci 27:5280–5290.
Sakowitz OW, Unterberg AW, Stover JF (2002) Neuronal activity determined by quantitative EEG and cortical microdialysis is increased following controlled cortical impact injury in rats. Acta Neurochir Suppl 81:221–223.
Salazar A, Jabbari B, Vance S, Grafman J, Amin D, Dillon J (1985) Epilepsy after penetrating head injury. I. Clinical correlates: a report of the Vietnam Head Injury Study. Neurology 35:1406–1414.
Salin P, Tseng G-F, Hoffman S, Parada I, Prince DA (1995) Axonal sprouting in layer V pyramidal neurons of chronically injured cerebral cortex. J Neurosci 15:8234–8245.
Sharpless SK (1969) Isolated and deafferented neurons: diffuse supersensitivity. In: Basic Mechanisms of the Epilepsies (Jasper H WA, and Pope A, ed), pp 329–348. Boston, MA: Little Brown.
Sharpless SK, Halpern LM (1962) The electrical excitability of chronically isolated cortex studied by means of permanently implanted electrodes. Electroencephalogr Clin Neurophysiol 14:244–255.

Shu Y, Hasenstaub A, McCormick DA (2003) Turning on and off recurrent balanced cortical activity. Nature 423:288–293.

Silva-Barrat C, Araneda S, Menini C, Champagnat J, Naquet R (1992) Burst generation in neocortical neurons after GABA withdrawal in the rat. J Neurophysiol 67:715–727.

Steriade M, Contreras D (1995) Relations between cortical and thalamic cellular events during transition from sleep patterns to paroxysmal activity. J Neurosci 15:623–642.

Steriade M, Contreras D (1998) Spike-wave complexes and fast components of cortically generated seizures. I. Role of neocortex and thalamus. J Neurophysiol 80:1439–1455.

Steriade M, Timofeev I, Grenier F (2001) Natural waking and sleep states: a view from inside neocortical neurons. J Neurophysiol 85:1969–1985.

Steriade M, Amzica F, Neckelmann D, Timofeev I (1998) Spike-wave complexes and fast components of cortically generated seizures. II. Extra- and intracellular patterns. J Neurophysiol 80:1456–1479.

Temkin NR, Haglund MM, Winn HR (1995) Causes, prevention, and treatment of post-traumatic epilepsy. New Horiz 3:518–522.

Temkin NR, Dikmen SS, Wilensky AJ, Keihm J, Chabal S, Winn HR (1990) A randomized, double-blind study of phenytoin for the prevention of post-traumatic seizures. N Engl J Med 323:497–502.

Temkin NR, Dikmen SS, Anderson GD, Wilensky AJ, Holmes MD, Cohen W, Newell DW, Nelson P, Awan A, Winn HR (1999) Valproate therapy for prevention of posttraumatic seizures: a randomized trial. J Neurosurg 91:593–600.

Thimm J, Mechler A, Lin H, Rhee S, Lal R (2005) Calcium-dependent open/closed conformations and interfacial energy maps of reconstituted hemichannels. J Biol Chem 280:10646–10654.

Timofeev I (2010) Pathophysiology of neocortical seizures. In: The Atlas of Epilepsies (Panayiotopoulos CP, ed), p in press. London: Springer-Verlag.

Timofeev I, Steriade M (2004) Neocortical seizures: initiation, development and cessation. Neuroscience 123:299–336.

Timofeev I, Bazhenov M (2005) Mechanisms of cortical trauma induced epileptogenesis and seizures. In: Recent Res. Devel. Physiol. (Pandalai SG, ed), pp 99–139. Kerala, India: Research Signpost.

Timofeev I, Bazhenov M, Sejnowski T, Steriade M (2002) Cortical hyperpolarization-activated depolarizing current takes part in the generation of focal paroxysmal activities. Proc Natl Acad Sci USA 99:9533–9537.

Timofeev I, Contreras D, Steriade M (1996) Synaptic responsiveness of cortical and thalamic neurones during various phases of slow sleep oscillation in cat. J Physiol 494:265–278.

Timofeev I, Grenier F, Steriade M (1998) Spike-wave complexes and fast components of cortically generated seizures. IV. Paroxysmal fast runs in cortical and thalamic neurons. J Neurophysiol 80:1495–1513.

Timofeev I, Grenier F, Steriade M (2000a) Impact of intrinsic properties and synaptic factors on the activity of neocortical networks in vivo. J Physiol (Paris) 94:343–355.

Timofeev I, Grenier F, Steriade M (2001) Disfacilitation and active inhibition in the neocortex during the natural sleep-wake cycle: An intracellular study. Proc Natl Acad Sci USA 98:1924–1929.

Timofeev I, Grenier F, Steriade M (2002) The role of chloride-dependent inhibition and the activity of fast-spiking neurons during cortical spike-wave seizures. Neuroscience 114:1115–1132.

Timofeev I, Grenier F, Steriade M (2004) Contribution of intrinsic neuronal factors in the generation of cortically driven electrographic seizures. J Neurophysiol 92:1133–1143.

Timofeev I, Bazhenov M, Avramescu S, Nita DA (2010) Posttraumatic epilepsy: the roles of synaptic plasticity. Neuroscientist 16:19–27.

Timofeev I, Grenier F, Bazhenov M, Sejnowski TJ, Steriade M (2000b) Origin of slow cortical oscillations in deafferented cortical slabs. Cereb Cortex 10:1185–1199.

Topolnik L, Steriade M, Timofeev I (2003a) Partial cortical deafferentation promotes development of paroxysmal activity. Cereb Cortex 13:883–893.

Topolnik L, Steriade M, Timofeev I (2003b) Hyperexcitability of intact neurons underlies acute development of trauma-related electrographic seizures in cats in vivo. Eur J Neurosci 18:486–496.

Traynelis SF, Dingledine R (1988) Potassium-induced spontaneous electrographic seizures in the rat hippocampal slice. J Neurophysiol 59:259–276.

Tseng G-F, Prince DA (1996) Structural and functional alterations in rat corticospinal neurons after axotomy. J Neurophysiol 75:248–267.

Turrigiano GG (1999) Homeostatic plasticity in neuronal networks: the more things change, the more they stay the same. Trends Neurosci 22:221–227.

Turrigiano GG, Leslie KR, Desai NS, Rutherford LC, Nelson SB (1998) Activity-dependent scaling of quantal amplitude in neocortical neurons. Nature 391:892–896.

Willmore LJ, Sypert GW, Munson JV, Hurd RW (1978) Chronic focal epileptiform discharges induced by injection of iron into rat and cat cortex. Science 200:1501–1503.

Zhang F, Wang LP, Boyden ES, Deisseroth K (2006) Channelrhodopsin-2 and optical control of excitable cells. Nat Methods 3:785–792.

Chapter 9
Cellular Mechanisms of Neuronal Cl^- Homeostasis and its Modulation by Neuronal Injury

A.J. Moorhouse and J. Nabekura

Abstract The function of the brain depends on the fine balance between inhibition and excitation of neurons. The polarity and amplitude of an inhibitory synaptic response depends on both the relative permeability of the receptor-channels involved, and also on the driving force for the permeant ions which is established by an array of active membrane transport proteins. At inhibitory synapses, the cation chloride co-transporters, including NKCC1 and KCC2, are particularly important in determining intracellular Cl^- concentrations and the efficacy of inhibitory synaptic responses. Recent experiments have shown that the expression and activity of these transporters can change, not only over development, but also in response to neuronal injury. In this article, we review the basic biophysical aspects and role of different transporters in neuronal Cl^- homeostasis and how the expression and function of these transporters, particularly KCC2, changes during development and in response to a wide range of models of neuronal injury. We also review recent data regarding the cellular and molecular mechanisms by which injury induced changes in KCC2 function may occur. Our review hopes to highlight the need for further investigation of these processes, to enable a greater understanding of developmental and pathological plasticity at inhibitory synapses, and to enable the potential development of novel therapeutic strategies to treat neuronal dysfunction.

Introduction

The generation and propagation of electrical and chemical signals between nerve cells in the brain depends critically on the co-ordinated activity of diverse neuronal membrane transport proteins. Different ion channels with distinct selectivities and

A.J. Moorhouse (✉)
Membrane Biophysics Laboratory, Department of Physiology, School of Medical Sciences, University of New South Wales, Sydney 2052, Australia
and
Division of Homeostatic Development, Department of Developmental Physiology, National Institute for Physiological Sciences, Okazaki 444-8585, Japan
e-mail: a.moorhouse@unsw.edu.au

M.A. Woodin and A. Maffei (eds.), *Inhibitory Synaptic Plasticity*,
DOI 10.1007/978-1-4419-6978-1_9,

gating mechanisms allow passive transmembrane movement of ions down their electrochemical gradients and the generation of various resting, synaptic, and action potentials. Of equal importance, is the primary and secondary active transporters that establish and maintain the ionic gradients necessary for the production of these electrical signals. Inhibitory synapses provide an excellent illustration of the close functional coupling between these elements and how alterations in the function of such active transporters can profoundly influence inhibitory synaptic transmission and brain physiology.

Contribution of Relative Ion Permeability to GABA and Glycine Synaptic Responses

γ-amino butyric acid (GABA) and glycine are the primary neurotransmitters responsible for fast signaling at inhibitory synapses in the mammalian brain. These neurotransmitters activate ionotropic receptors predominantly permeable to anions. Although Cl^- is the physiological ion with the highest permeability through these ion channels, HCO_3^- ions can also pass through these channels, with a permeability relative to Cl^- of approximately 0.2–0.4 (Kaila 1994; Farrant and Kaila 2007). As the electrochemical driving force for HCO_3^- in resting neurons will be for an efflux under virtually all physiological conditions (e.g., Figure 5 of Payne et al. 2003), the HCO_3^- permeability will enhance depolarizing GABA/glycine responses, or reduce hyperpolarizing responses. Some modest but potentially significant cation permeability may also occur through native GABA and glycine receptors, as reported for some recombinant receptor-channels (e.g., $P_{Cl^-} : P_{Na^+} < 20$; Carland et al. 2004, 2009; Sugiharto et al. 2008), which would also tend to favour depolarizations. Nevertherless, Cl^- homeostasis is the major factor that determines the presence and polarity of membrane currents induced by activation of these ionotropic receptors, and this is determined primarily by the expression and function of the array of different active transporters present in neurons.

Active Transporters Contributing to Cl^- Homeostasis in Neurons

The conductance change associated with GABA or glycine activation of ionotropic receptors at inhibitory synapses at a typical resting membrane potential (V_m, ~70 mV) may elicit a depolarization (Cl^- efflux) or a hyperpolarization (Cl^- influx) depending on the intracellular Cl^- concentration ($[Cl^-]_i$). The polarity of the synaptic responses does not necessarily directly correlate to excitation and inhibition of the postsynaptic neuron, and the conductance change itself will result in significant shunting inhibition of any excitatory inputs (Farrant and Kaila 2007; Blaesse et al. 2009). To elicit a membrane potential change, the permeant ions must not be at equilibrium at rest, and hence must be actively transported across the cell membrane. The major transporters influencing $[Cl^-]_i$ belong to the SLC12 family of secondary-active membrane transporters (Payne et al. 2003; Hebert et al. 2004; Blaesse et al. 2009). These are electroneutral cation-chloride cotransporters (CCCs) comprising the different isoforms of the $Na^+K^+Cl^-$ cotransporters (SLC12A1-2; NKCC2-1), the Na^+Cl^- co-transporter (SLC12A3, NCC), the K^+Cl^- cotransporters (SLC12A4-7; KCC1-4) and the poorly characterized CIP1 (CCC-interacting protein 1; SLC12A8) and CCC (SLC12A9) transporters. The NKCC2 and NCC transporters are exclusively expressed in the kidney, and hence the identified neuronal population, including their splice variants, include NKCC1a, 1b; KCC1, KCC2a, 2b; KCC3a, 3b, 3c; KCC4. Under typical resting conditions (Fig. 9.1a), NKCC1 couples the electrochemical gradient for Na^+ influx to Cl^- (and K^+) influx, while the KCCs couple the electrochemical gradient for K^+ efflux to Cl^- efflux. If present and active, NKCCs will therefore increase $[Cl^-]_i$ and hence will favour a more depolarized Cl^- equilibrium potential and GABA/glycine response, while KCCs will decrease $[Cl^-]_i$ and hence favour a more hyperpolarized Cl^- equilibrium potential and GABA/glycine response.

Neuronal Cl^- homeostasis is also affected by activity of the SLC4a and SLC26a Na^+-dependent and Na^+-independent anion-exchangers. These transporters couple HCO^{3-} efflux down its electrochemical gradient to Cl^- influx and hence would facilitate a depolarizing GABA/glycine response in neurons (e.g., Nickell et al. 2007; Fig. 9.1a). The HCO^{3-} driving force is coupled to intracellular pH and carbonic anhydrase activity, and HCO^{3-} changes, and may play particularly important roles in GABAergic depolarizations (disinhibition) observed during intense activity (Rivera et al. 2005). In addition, a primary active transporter, a Cl-ATPase has been reported in mammalian neurons (Inagaki et al. 2001), although its precise characteristics in the plasma membrane and its potential contribution to transmission at inhibitory synapses in the brain is less clear. While many additional anion channels, activated by voltage, cell swelling, intracellular Ca^{2+} and other mechanisms exist in neurons, they are passive transporters that do not play a role in determining Cl^- flux at inhibitory synapses (although volume changes can result in changes in $[Cl^-]_i$). In summary, a great diversity of coupling between physiologically relevant ions and transporters can regulate Cl^- homeostasis, providing many potential mechanisms by which Cl^- homeostasis can be regulated under different developmental, physiological and pathological conditions. The remainder of this article will focus on changes in KCC2 and NKCC1 function and expression and their influence on inhibitory synaptic transmission.

Diversity of Distribution and Functional Expression of NKCC1 and KCC2

The neuronal $[Cl^-]_i$ may primarily depend on the relative activities of NKCC1 and KCCs in the plasma membrane, which in turn depends on both the expression levels and the activity of these transporters. To date, most studies have focused on

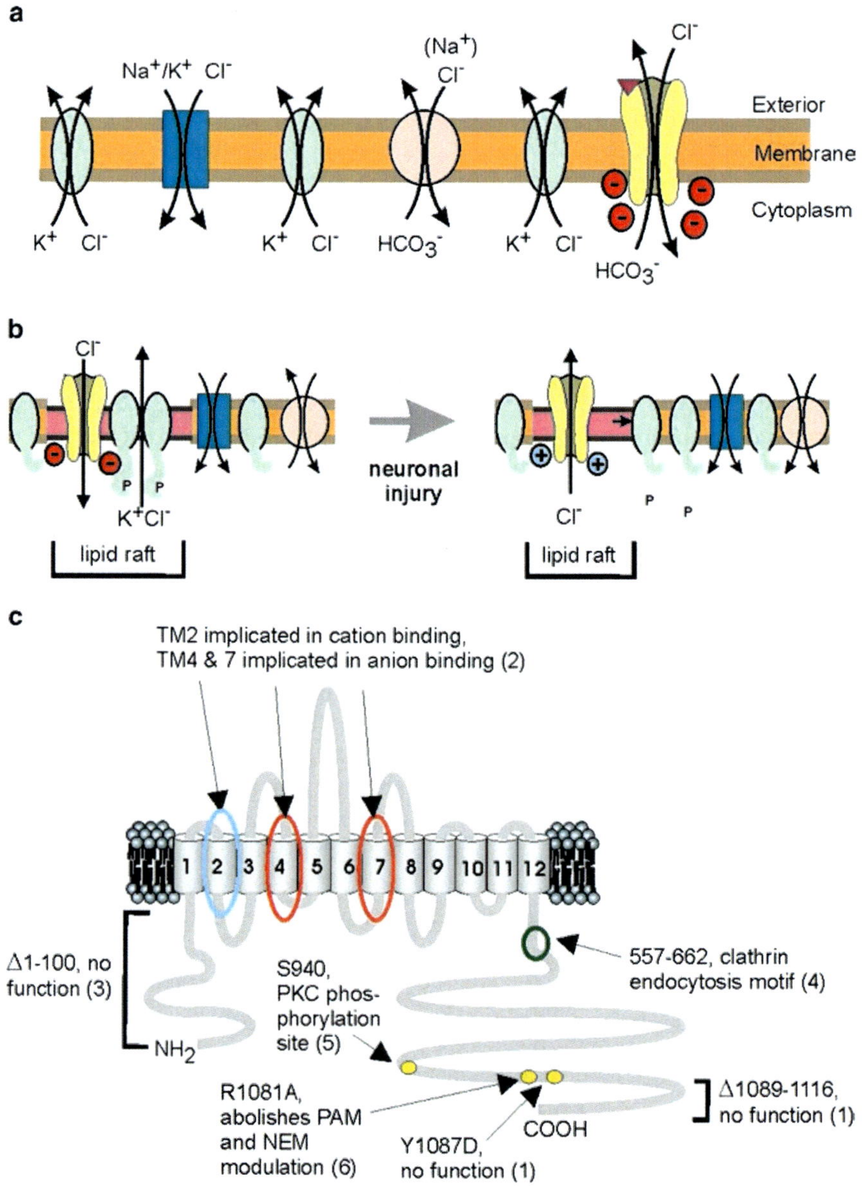

Fig. 9.1 (**a**) Schematic illustration of the known secondary active transporters involved in neuronal Cl^- homeostasis and the direction of ion transport under normal resting conditions, including (from *left* to *right*); KCC (*light green oval*, K^+ and Cl^- efflux), NKCC (*blue square*, Na^+, K^+ and Cl^- influx), and the Na^+ dependent and independent anion exchangers (*pink circle*, Na^+, Cl influx, HCO_3^- efflux). An ionotropic GABA receptor is also shown, activated by GABA (*purple triangle*) and permeant to Cl^- (influx) and HCO_3^- (efflux). The dominance of KCC2 to Cl^- homeostasis in many mature neurons under normal conditions gives rise to GABA and glycine responses which typically hyperpolarize the postsynaptic membrane (indicated by intracellular –'ve charges in orange).

KCC2 and less is known about the contributions of other KCCs to neuronal $[Cl^-]_i$. Initial characterisation of KCC1, KCC3 and KCC4 in *Xenopus* oocytes indicated that transport was only observed in response to cell swelling, although basal transport is seen in mammalian (HEK293) cells (Gamba 2005). The expression of NKCC1 and KCC2 differs markedly throughout the brain, with different patterns of distribution at both the regional and sub-cellular level, and different expression levels throughout development. NKCC1 expression is apparent in embryonic and neonatal neurons prior to KCC2 expression and is responsible for depolarizing GABA and glycine responses in early developmental (embryonic, neonatal) periods (Clayton et al. 1998; Kakazu et al. 1999; Yamada et al. 2004). A developmental increase (with a variable time-course) in expression levels of KCC2 is primarily responsible for a hyperpolarizing GABA and glycine response (where present) in mature neuronal somata (Rivera et al. 1999; Kakazu et al. 1999). The KCC2b variant is responsible for this developmental upregulation and is the dominant adult KCC2 isoform (Uvarov et al. 2007; 2009). There is debate as to the presence and extent of a parallel downregulation of NKCC1 (Clayton et al. 1998; Kakazu et al. 1999; Yamada et al. 2004; Blaesse et al. 2009). However KCC2 remains absent or very low in some mature neurons where $[Cl^-]_i$ homeostasis is dominated by NKCC1 and/or by anion-exchangers or other mechanisms (e.g. DRGs, Rocha-González et al. 2008; substantia nigra dopaminergic neurons, Gulácsi et al. 2003; olfactory receptor neurons, Nickell et al. 2007; hypothalamic GnRH neurons, Watanabe et al. 2009a). Sub-cellular differences may also exist, for example, presynaptic terminals in isolated hypothalamic and hippocampal neurons respond to GABA with depolarization due to a high resting NKCC1 activity (Jang et al. 2001, 2006).

Expression levels, however, may not be directly correlated to transport activity. KCC2, for example, seems to also have a morphogenic function, independent of transport, in promoting maturation of development of dendritic spines and associated excitatory synapses (Li et al. 2007). In developing rat auditory brainstem

Fig. 9.1 (continued) (**b**) Schematic model of membrane Cl^- transport and ionotropic GABA receptor responses during control conditions (*left*) and in response to neuronal injury induced in cultured hippocampal neurons by epileptic activity (zero external Mg^{2+}) or oxidative stress (H_2O_2), and as based on Wake et al. (2007) and Watanabe et al. (2009b). The membrane transport proteins use the same colour code as shown in *panel A*. Under control conditions, KCC2 is functional, phosphorylated at tyrosine 1089 in the intracellular carboxy-tail, and associated with oligomers and lipid rafts, and hence GABA induces Cl^- influx and a hyperpolarization. Following neuronal injury, KCC2 looses its tyrosine phosphorylation, its association with lipid rafts, and its transport function, and GABA induces Cl^- efflux and depolarization. (**c**) Schematic diagram of KCC2 primary structure, indicating the membrane topology with the proposed 12 transmembrane (TM) domains (1–12) and the approximate relative length of amino acid chains in the amino (NH_3) and carboxy (COOH) termini, and for the loops connecting the TM helices (from Song et al. 2002; Gamba 2005, note TM 5 is absent in Fig. 9.1 of Song et al. 2002). Some residue mutations and/or deletions shown to have major effects on KCC2 transport function are indicated. References quoted are: (1) Watanabe et al. 2009b, (2) Gamba 2005 (based on analogy to NKCC), (3) Li et al. 2007, (4) Zhao et al. 2008, (5) Lee et al. 2007, (6) Garbarini and Delpire 2008 (and other) neurons (Blaesse et al. 2006), and in early cultured cortical neurons (e.g., Khirug et al. 2005), KCC2 is present but its transport activity completely absent. The activity of NKCC1 and KCC2 is regulated by different kinases and phosphatases. A consensus derived mostly from non-neuronal cells is that phosphorylation activates NKCC and inhibits KCC activity (Adragna et al. 2004; Gamba 2005) although this is an over-simplification for KCC2 (e.g., below) and more work is needed to identify the precise enzymes involved in acute and chronic regulation of Cl^- transporters in the brain, and the conditions under which this may occur in vivo. Transport activity also depends on substrate availability, and this is particularly important for KCC2 which is not only close to its thermodynamic equilibrium under resting conditions (as $E_{K^+} \sim E_{Cl^-}$), but also operates with substrate affinities close to the physiological range for $[Cl^-]_i$ and $[K^+]_o$ (Gamba 2005). Hence both the rate and direction of transport will be particularly sensitive to changes in $[Cl^-]_i$ and $[K^+]_o$, and reversal of transport as extracellular $[K^+]$ is increased has been observed (Kakazu et al. 2000). Similarly, an enhanced intracellular $[Cl^-]$ load would be expected to enhance KCC2 efflux transport and increase extracellular K^+, as observed during intense $GABA_A$ receptor activation (Barolet and Morris 1991; see also Payne et al. 2003; Rivera et al. 2005).

Changes in Cl^- Homeostasis Is a Common Feature of Diverse Neuronal Injuries

As indicated above, in immature central neurons, $[Cl^-]_i$ is relatively high and GABA and glycine elicit depolarizing synaptic responses (Ben-Ari 2002). With maturation and increasing KCC2b expression, and in some cases coupled to reduced NKCC21 expression, there is a decrease in $[Cl^-]_i$ and the GABA/glycine response becomes hyperpolarizing (Rivera et al. 1999; Kakazu et al. 1999). In addition to this developmental change in GABA response polarity, differences in KCC2 expression levels in distinct nuclei correlate to their ability to maintain Cl^- homeostasis during repetitive stimulation or sustained depolarizations (Ueno et al. 2002; Zhu et al. 2005). A reversal of this developmental pattern, namely a reduced expression and function of KCC2 and a resultant increase in $[Cl^-]_i$, occurs in the mature nervous system in response to a variety of pathological insults (Table 9.1), including axotomy, epilepsy, ischemia, blunt trauma and neuropathic pain (see also Kahle et al. 2008; Blaesse et al. 2009). This can also be replicated in vitro, and examples of both in vivo and in vitro models of neuronal injury that result in functional KCC2 downregulation are shown in Table 9.1. In many cases, the injury is not only associated with reduced GABA hyperpolarization, but a true reversion back to a depolarizing GABA response is apparent, indicating Cl^- efflux which is only possible if active Cl^- influx via NKCC1 or another transport mechanism are maintained in injured mature neurons (e.g. Shulga et al. 2008). In fact, NKCC expression and function may be increased in response to different injury models (e.g., seizures, Okabe et al. 2002; ischemia, Pond et al. 2006). In some injury models, KCC2

Table 9.1 Downregulation of KCC2 expression in response to various neuronal pathologies in vitro and in vivo

Neuronal injury	Description of injury model	Effect on KCC2 expression & GABA responses	References
Axotomy in vivo	Cutting vagus or facial nerves, or corticospinal tract, KCC2 assayed in axotomized soma	Reduced KCC2 expression at 8[a] h – 2 days after axotomy, +'ve E_{GABA} shift	Nabekura et al. (2002), Toyoda et al. (2003), Shulga et al. (2008)
Axotomy in vitro	Cutting CA3 Schaffer collateral afferents, assay KCC2 in axotomized soma	KCC2 protein & mRNA decreased (10–20%), GABA induces depolarizations	Shulga et al. (2008)
Neuropathic pain in vivo	Chronic sciatic nerve cuff application, noicicpetive sensitization (1–2 weeks later) confirmed with behavioural testing	Reduced KCC2 expression in dorsal horn neurons (1–2 weeks)[a] with +'ve E_{GABA} shift, and GABA induced depolarizations	Coull et al. (2003, 2005)
Ischemia in vivo	15 min transient forebrain ischemia (four vessel occlusion) in adult rats	Initial increase in KCC2 expression (6 h) followed by decreased expression as neurons die	Papp et al. (2008)
Ischemia in vitro	7 min perfusion of adult rat hippocampal slices with anoxic, 0 glucose media	KCC2 expression decreases 1[a]–2 h after reoxygenation, $[Cl^-]_i$ increases	Galeffi et al. (2004)
Epilepsy in vivo	Subiculum tissues excised from epileptic humans	Subsets of reduced KCC2 expression, +'ve E_{GABA} shift	Huberfeld et al. (2007)
Siezure in vitro	Application of excitable solutions (0 Mg^{2+}) to hippocampal neurons	KCC2 protein decreased within 1[a]–3 h, +'ve E_{GABA} shift	Rivera et al. (2004), Wake et al. (2007)
Oxidative stress in vitro	Application of H_2O_2 to cultured hippocampal neurons	KCC2 function decreased <1[a] h, +'ve E_{GABA} shift	Wake et al. (2007)

[a] This was the earliest time point examined in the studies

expression levels have been shown to increase. Following forebrain ischemia in vivo, for example, it was suggested that increased KCC2 may be an adaptive response to minimize subsequent damage (Papp et al. 2008), although functional activity wasn't measured.

The changes seen in these pathological conditions resemble a return of neuronal $[Cl^-]_i$ homeostasis to a more immature phenotype. Given the role of depolarizing GABA responses (and subsequent Ca^{2+} influxes) in neuronal migration, synapse formation and in patterns of excitatory activity important for establishing neuronal circuits (Represa and Ben-Ari; 2005), it has been speculated that this temporary restoration of the immature state may be important for adaptations required to re-establish appropriate neuronal connections and functions (Payne et al. 2003;

Toyoda et al. 2003), and perhaps associated with a similar "critical period" for recovery (Shulga et al. 2008). However direct evidence for this hypothesis is lacking, and rapid KCC2 downregulation and depolarizing GABA depolarizing responses can contribute to neuronal injury via excitotoxic mechanisms (e.g., Wake et al. 2007) and reduced KCC2 function confers increased susceptibility to seizure and ischemic damage (Tornberg et al. 2005; Papp et al. 2008). Restoring Cl^- homeostasis by enhancing KCC2 function, or inhibiting NKCC1 has potential as a therapeutic strategy to treat epilepsy and neuropathic pain, and to reduce stroke-induced damage, and clinical trials with the NKCC1 inhibitor, bumetanide, to treat neonatal epilepsies are currently underway (Kahle et al. 2008).

Molecular and Cellular Mechanisms Mediating Injury-Induced KCC2 Downregulation

The transport function of KCC2 can be regulated by changes in expression levels and/or by changes in the activity of expressed KCC2. Numerous studies (Table 9.1) have reported reduced expression levels of KCC2 after injury, although in only a few of these models have the potential mechanisms linking injury and downregulation been investigated. Blocking the actions of the brain-derived neurotrophic factor (BDNF), prevents the reduction of KCC2 expression which occurs in isolated hippocampal slices in response to experimental induction of inter-ictal (epileptic) activity by incubation with 4-AP or 0 Mg^{2+} (Rivera et al. 2002, 2004). ATP induced release of BDNF from microglia results in reduced KCC2 levels in dorsal horn neurons that is responsible for the resultant allodynia and neuropathic pain following chronic sciatic nerve compression (Coull et al. 2003, 2005). The turnover rate of KCC2 at the plasma membrane is quite rapid, τ, ~ 20 min in cultured hippocampal neurons (Rivera et al. 2004) and even faster in HEK-293 cells (Lee et al. 2007), and hence changes in surface protein expression may mediate changes in function observed on relatively rapid time scales. Direct phosphorylation of KCC2 by PKC or by tyrosine kinase can result in relatively rapid changes (~10–<60 min) in surface expression or distribution, that result in altered cellular KCC2 transport, independent of changes in total KCC2 protein levels (Lee et al. 2007; Watanabe et al. 2009b). A number of studies have reported changes in transport function due to changes in KCC2 phosphorylation state (Gamba 2005), although it is not always clear if the effects involve direct changes in the phsophorylation state of KCC2 itself, or are mediated via an intermediate protein. Pull-down and co-expression studies have shown that KCC2 function can also be increased by direct interactions with accessory proteins, including brain creatinine kinase (Inoue et al. 2006), SLC12A8/CIP (Wenz et al. 2009) and Protein Associated with Myc (PAM; Garbarini and Delpire 2008) however the relevance of these interactions to normal and pathological changes in KCC2 is unclear.

In addition to the role of BDNF in injury-induced KCC2 protein expression downregulation described above, injury-induced changes in KCC2 function, independent of any changes in expression levels, have also been reported. For example, in response to oxidative stress (H_2O_2 application), or induction of inter-ictal activity (0 Mg^{2+}), an early phase (<1 h) of downregulation of KCC2 activity occurs by loss of direct phosphorylation of a tyrosine residue (Y1089) in the intracellular carboxy tail of KCC2 (Fig. 9.1b, c; Wake et al. 2007; Watanabe et al. 2009b). Tyrosine phosphorylation of this residue, and indeed the presence of the carboxy tail distal to this residue, is critical for KCC2 transport function, and is associated with the localization of functional KCC2 in lipid raft microdomains (Fig. 9.1b, c; Watanabe et al. 2009b). Liberation of intracellular Zn^{2+} was recently shown to mediate downregulation of KCC2 activity following oxygen and glucose deprivation of cultured cortical neurons, again this was independent of any changes in surface expression levels or KCC2 oligomerisation state (Hershfinkel et al. 2009). Further studies on the mechanisms responsible for injury induced changes in KCC2, as well as those to expand the limited structure-function knowledge of KCC2 (Fig. 9.1c), are clearly warranted.

Concluding Remarks

The efficacy of inhibitory synaptic transmission depends critically not only on properties of the inhibitory receptor-channels responding to GABA and glycine and the transporters that determine the neurotransmitter time-course, but also on the different secondary active transporters that determine $[Cl^-]_i$ homeostasis. Principle amongst these transporters are NKCC1 and KCC2. Different relative expression patterns of these transporters can determine whether GABA and glycine induce a depolarization or hyperpolarization at synapses, and developmental changes in their relative expression mediates developmental changes the polarity of GABAergic responses. More recent data has shown that the expression and activity of these transporters can also change in response to neuronal injury. In particular, injury induced decreases in the surface expression and activity of KCC2, and a concomittent conversion of GABA and glycine responses towards depolarization and excitation, are common to a number of different models of injury. The challenge for neuroscientists is to elucidate the functional consequences of this change in GABAergic signaling at inhibitory synapses, to uncover the cellular and molecular mechanisms mediating the change in KCC2 function, and finally to apply this knowledge to novel approaches to treat these injuries to the brain.

Acknowledgements We thank Miho Watanabe for helpful comments.

References

Adragna NC, Fulvio MD, Lauf PK (2004) Regulation of K-Cl cotransport: from function to genes. J Membr Biol 201:109–137.

Ben-Ari Y (2002) Excitatory actions of GABA during development: the nature of the nurture. Nat Rev Neurosci 3:728–739.

Blaesse P, Guillemin I, Schindler J, Schweizer M, Delpire E, Khiroug L, Friauf E, Nothwang HG (2006) Oligomerization of KCC2 correlates with development of inhibitory neurotransmission. J Neurosci 26:10407–10419.

Blaesse P, Airaksinen MS, Rivera C, Kaila K (2009) Cation-chloride cotransporters and neuronal function. Neuron 61:820–838.

Barolet AW, Morris ME (1991) Changes in extracellular K^+ evoked by GABA, THIP and baclofen in the guinea-pig hippocampal slice. Exp Brain Res 84:591–598.

Carland JE, Moorhouse AJ, Barry PH, Johnston GA, Chebib M (2004) Charged residues at the 2′ position of human GABAC rho 1 receptors invert ion selectivity and influence open state probability. J Biol Chem 279:54153–54160.

Carland JE, Cooper MA, Sugiharto S, Jeong HJ, Lewis TM, Barry PH, Peters JA, Lambert JJ, Moorhouse AJ (2009) Characterization of the effects of charged residues in the intracellular loop on ion permeation in alpha1 glycine receptor channels. J Biol Chem 284:2023–2030.

Clayton GH, Owens GC, Wolff JS, Smith RL (1998) Ontogeny of cation-Cl- cotransporter expression in rat neocortex. Dev Brain Res 109:281–292.

Coull JA, Boudreau D, Bachand K, Prescott SA, Nault F, Sik A, De Koninck P, De Koninck Y (2003) Trans-synaptic shift in anion gradient in spinal lamina I neurons as a mechanism of neuropathic pain. Nature 424:938–942.

Coull JA, Beggs S, Boudreau D, Boivin D, Tsuda M, Inoue K, Gravel C, Salter MW, De Koninck Y (2005) BDNF from microglia causes the shift in neuronal anion gradient underlying neuropathic pain. Nature 438:1017–1021.

Farrant M, Kaila K (2007) The cellular, molecular and ionic basis of GABA(A) receptor signalling. Prog Brain Res 160:59–87.

Galeffi F, Sah R, Pond BB, George A, Schwartz-Bloom RD (2004) Changes in intracellular chloride after oxygen-glucose deprivation of the adult hippocampal slice: effect of diazepam. J Neurosci 24:4478–4488.

Gamba G (2005) Molecular physiology and pathophysiology of electroneutral cation-chloride cotransporters. Physiol Rev 85:423–493.

Garbarini N, Delpire E (2008) The RCC1 domain of protein associated with Myc (PAM) interacts with and regulates KCC2. Cell Physiol Biochem 22:31–44.

Gulácsi A, Lee CR, Sík A, Viitanen T, Kaila K, Tepper JM, Freund TF (2003) Cell type-specific differences in chloride-regulatory mechanisms and GABA(A) receptor-mediated inhibition in rat substantia nigra. J Neurosci 23:8237–8246.

Hebert SC, Mount DB, Gamba G (2004) Molecular physiology of cation-coupled Cl- cotransport: the SLC12 family. Pflugers Arch 447:580–593.

Hershfinkel M, Kandler K, Knoch ME, Dagan-Rabin M, Aras MA, Abramovitch-Dahan C, Sekler I, Aizenman E (2009) Intracellular zinc inhibits KCC2 transporter activity. Nat Neurosci 12:725–727.

Huberfeld G, Wittner L, Clemenceau S, Baulac M, Kaila K, Miles R, Rivera C (2007) Perturbed chloride homeostasis and GABAergic signaling in human temporal lobe epilepsy. J Neurosci 27:9866–9873.

Inagaki C, Hattori N, Kitagawa K, Zeng XT, Yagyu K (2001) $Cl(^-)$-ATPase in rat brain and kidney. J Exp Zool 289:224–231.

Inoue K, Yamada J, Ueno S, Fukuda A (2006) Brain-type creatine kinase activates neuron-specific K^+-Cl^- co-transporter KCC2. J Neurochem 96:598–608.

Jang IS, Jeong HJ, Akaike N (2001) Contribution of the Na-K-Cl cotransporter on GABA(A) receptor-mediated presynaptic depolarization in excitatory nerve terminals. J Neurosci 21:5962–5972.

Jang IS, Nakamura M, Ito Y, Akaike N (2006) Presynaptic GABAA receptors facilitate spontaneous glutamate release from presynaptic terminals on mechanically dissociated rat CA3 pyramidal neurons. Neuroscience 138:25–35.

Kaila K (1994) Ionic basis of GABAA receptor channel function in the nervous system. Prog Neurobiol 42:489–537.

Kakazu Y, Akaike N, Komiyama S, Nabekura J (1999) Regulation of intracellular chloride by cotransporters in developing lateral superior olive neurons. J Neurosci 19:2843–2851.

Kakazu Y, Uchida S, Nakagawa T, Akaike N, Nabekura J (2000) Reversibility and cation selectivity of the K(+)-Cl(−) cotransport in rat central neurons. J Neurophysiol 84:281–288.
Kahle KT, Staley KJ, Nahed BV, Gamba G, Hebert SC, Lifton RP, Mount DB (2008) Roles of the cation-chloride cotransporters in neurological disease. Nat Clin Pract Neurol 4:490–503.
Khirug S, Huttu K, Ludwig A, Smirnov S, Voipio J, Rivera C, Kaila K, Khiroug L (2005) Distinct properties of functional KCC2 expression in immature mouse hippocampal neurons in culture and in acute slices. Eur J Neurosci 21:899–904.
Lee HH, Walker JA, Williams JR, Goodier RJ, Payne JA, Moss SJ (2007) Direct protein kinase C-dependent phosphorylation regulates the cell surface stability and activity of the potassium chloride cotransporter KCC2. J Biol Chem 282:29777–29784.
Li H, Khirug S, Cai C, Ludwig A, Blaesse P, Kolikova J, Afzalov R, Coleman SK, Lauri S, Airaksinen MS, Keinänen K, Khiroug L, Saarma M, Kaila K, Rivera C (2007) KCC2 interacts with the dendritic cytoskeleton to promote spine development. Neuron 56:1019–1033.
Nabekura J, Ueno T, Okabe A, Furuta A, Iwaki T, Shimizu-Okabe C, Fukuda A, Akaike N (2002) Reduction of KCC2 expression and GABAA receptor-mediated excitation after in vivo axonal injury. J Neurosci 22:4412–4417.
Nickell WT, Kleene NK, Kleene SJ (2007) Mechanisms of neuronal chloride accumulation in intact mouse olfactory epithelium. J Physiol 583:1005–1020.
Okabe A, Ohno K, Toyoda H, Yokokura M, Sato K, Fukuda A (2002) Amygdala kindling induces upregulation of mRNA for NKCC1, a Na(+), K(+)-2Cl(−) cotransporter, in the rat piriform cortex. Neurosci Res 44:225–229.
Papp E, Rivera C, Kaila K, Freund TF (2008) Relationship between neuronal vulnerability and potassium-chloride cotransporter 2 immunoreactivity in hippocampus following transient forebrain ischemia. Neuroscience 154:677–689.
Payne JA, Rivera C, Voipio J, Kaika K (2003) Cation-chloride co-transporters in neuronal communication, development and trauma. Trends Neurosci 26:199–206.
Pond BB, Berglund K, Kuner T, Feng G, Augustine GJ, Schwartz-Bloom RD (2006) The chloride transporter Na(+)-K(+)-Cl^- cotransporter isoform-1 contributes to intracellular chloride increases after in vitro ischemia. J Neurosci 26:1396–1406.
Represa A, Ben-Ari Y (2005) Trophic actions of GABA on neuronal development. Trends Neurosci 28:278–283.
Rivera C, Voipio J, Payne JA, Ruusuvuori E, Lahtinen H, Lamsa K, Pirvola U, Saarma M, Kaila K (1999) The K^+/Cl^- co-transporter KCC2 renders GABA hyperpolarizing during neuronal maturation. Nature 397:251–255.
Rivera C, Li H, Thomas-Crusells J, Lahtinen H, Viitanen T, Nanobashvili A, Kokaia Z, Airaksinen MS, Voipio J, Kaila K, Saarma M (2002) BDNF-induced TrkB activation down-regulates the K^+-Cl^- cotransporter KCC2 and impairs neuronal Cl- extrusion. J Cell Biol 159:747–752.
Rivera C, Voipio J, Thomas-Crusells J, Li H, Emri Z, Sipila S, Payne JA, Minichiello L, Saarma M, Kaila K (2004) Mechanism of activity-dependent downregulation of the neuron-specific K-Cl cotransporter KCC2. J Neurosci 24:4683–4691.
Rivera C, Voipio J, Kaila K (2005) Two developmental switches in GABAergic signalling: the K^+-Cl^- cotransporter KCC2 and carbonic anhydrase CAVII. J Physiol 562:27–36.
Rocha-González HI, Mao S, Alvarez-Leefmans FJ (2008) Na^+,K^+,$2Cl^-$ cotransport and intracellular chloride regulation in rat primary sensory neurons: thermodynamic and kinetic aspects. J Neurophysiol 100:169–184.
Shulga A, Thomas-Crusells J, Sigl T, Blaesse A, Mestres P, Meyer M, Yan Q, Kaila K, Saarma M, Rivera C, Giehl KM (2008) Posttraumatic GABA(A)-mediated $[Ca2^+]$i increase is essential for the induction of brain-derived neurotrophic factor-dependent survival of mature central neurons. J Neurosci 28:6996–7005.
Song L, Mercado A, Vázquez N, Xie Q, Desai R, George AL Jr, Gamba G, Mount DB (2002) Molecular, functional, and genomic characterization of human KCC2, the neuronal K-Cl cotransporter. Mol Brain Res 103:91–105.
Sugiharto S, Lewis TM, Moorhouse AJ, Schofield PR, Barry PH (2008) Anion-cation permeability correlates with hydrated counterion size in glycine receptor channels. Biophys J 95:4698–4715.

Tornberg J, Voikar V, Savilahti H, Rauvala H, Airaksinen MS (2005) Behavioural phenotypes of hypomorphic KCC2-deficient mice. Eur J Neurosci 21:1327–1337.
Toyoda H, Ohno K, Yamada J, Ikeda M, Okabe A, Sato K, Hashimoto K, Fukuda A (2003) Induction of NMDA and GABAA receptor-mediated Ca2+ oscillations with KCC2 mRNA downregulation in injured facial motoneurons. J Neurophysiol 89:1353–1362.
Ueno T, Okabe A, Akaike N, Fukuda A, Nabekura J (2002) Diversity of neuron-specific K^+-Cl^- cotransporter expression and inhibitory postsynaptic potential depression in rat motoneurons. J Biol Chem 277:4945–4950.
Uvarov P, Ludwig A, Markkanen M, Pruunsild P, Kaila K, Delpire E, Timmusk T, Rivera C, Airaksinen MS (2007) A novel N-terminal isoform of the neuron-specific K-Cl cotransporter KCC2. J Biol Chem 282:30570–30576.
Uvarov P, Ludwig A, Markkanen M, Soni S, Hübner CA, Rivera C, Airaksinen MS (2009) Coexpression and heteromerization of two neuronal K-Cl cotransporter isoforms in neonatal brain. J Biol Chem 284:13696–13704.
Wake H, Watanabe M, Moorhouse AJ, Kanematsu K, Horibe S, Matsukawa N, Asai K, Ojika K, Hirata M, Nabekura J (2007) Early changes in KCC2 phosphorylation in response to neuronal stress results in functional downregulation. J Neurosci 27:1642–1650.
Watanabe M, Sakuma Y, Kato M (2009a) GABAA receptors mediate excitation in adult Rat GnRH neurons. Biol Reprod 81:327–332.
Watanabe M, Wake H, Moorhouse AJ, Nabekura J (2009b) Clustering of neuronal K^+-Cl^- cotransporters in lipid rafts by tyrosine phosphorylation. J Biol Chem 284:27980–27988.
Wenz M, Hartmann AM, Friauf E, Nothwang HG (2009) CIP1 is an activator of the K^+-Cl^- cotransporter KCC2. Biochem Biophys Res Commun 381:388–392.
Yamada J, Okabe A, Toyoda H, Kilb W, Luhmann HJ, Fukuda A (2004) Cl- uptake promoting depolarizing GABA actions in immature rat neocortical neurones is mediated by NKCC1. J Physiol 557:829–841.
Zhao B, Wong AY, Murshid A, Bowie D, Presley JF, Bedford FK (2008) Identification of a novel di-leucine motif mediating K(+)/Cl(−) cotransporter KCC2 constitutive endocytosis. Cell Signal 20:1769–1779.
Zhu L, Lovinger D, Delpire E (2005) Cortical neurons lacking KCC2 expression show impaired regulation of intracellular chloride. J Neurophysiol 93:1557–1568.

Part IV
Learning and Memory

Chapter 10
Activity-Dependent Inhibitory Synaptic Plasticity Mediated by Chloride Regulation

Trevor Balena, Brooke A. Acton, and Melanie A. Woodin

Abstract Synaptic plasticity is the ability of synapses to change their strength in response to either specific patterns of neuronal activity or the presence of certain chemicals. While the majority of research in this area has focused on excitatory glutamatergic synapses, synapses mediated by the neurotransmitter GABA have been receiving increasing attention. $GABA_A$-mediated synaptic transmission is primarily due to a flux of chloride across the membrane, and accounts for the majority of fast inhibitory synaptic transmission in the mature brain. GABAergic transmission transitions from excitatory to inhibitory during nervous system development due to changes in the expression of key cation-chloride cotransporters that determine the level of neuronal chloride. Recent studies have demonstrated that activity-dependent GABAergic synaptic plasticity can be induced in the embryonic, early postnatal and mature nervous systems. In this review, we will summarize recent work which explores activity-dependent inhibitory synaptic plasticity that results from changes in cation-chloride cotransporter regulation or expression.

Introduction

Synaptic plasticity studies have mainly focused on glutamate due to its prevalence in the central nervous system (CNS). Early work demonstrated that excitatory glutamatergic synapses undergo activity-dependent long-term potentiation (LTP) (Bliss and Lomo 1973; Malenka and Bear 2004). Activity-induced synaptic plasticity provides a cellular basis for the refinement of neuronal connections (Katz and Shatz 1996), and is thought to be crucial for memory formation in the hippocampus (Bliss and Collingridge 1993; Martin et al. 2000). GABAergic (γ-aminobutyric acid) synapses have also been shown to undergo activity-dependent synaptic

M.A. Woodin (✉)
Department of Cell & Systems Biology, University of Toronto, 25 Harbord Street, Toronto, ON M5S 3G5, Canada
e-mail: m.woodin@utoronto.ca

M.A. Woodin and A. Maffei (eds.), *Inhibitory Synaptic Plasticity*,
DOI 10.1007/978-1-4419-6978-1_10,

plasticity and be involved in learning and memory (Gaiarsa et al. 2002; Ramirez et al. 2005; Berlau and McGaugh 2006; Lee et al. 2006; Ben-Ari et al. 2007). Many forms of activity-dependent GABAergic plasticity are due to changes in synaptic conductance (Komatsu and Iwakiri 1993; McLean et al. 1996; Aizenman et al. 1998; Caillard et al. 1999, 2000; Gaiarsa et al. 2002; Lien et al. 2006; Ben-Ari et al. 2007). However because ionotropic $GABA_ARs$ are permeable to Cl^-, the strength and polarity of $GABA_A$-mediated inhibitory synaptic transmission are also determined by the reversal potential for GABA (E_{GABA}) (Kaila 1994; Farrant and Kaila 2007). This review focuses on how physiologically normal levels of neuronal activity[1] regulate neuronal Cl^-, resulting in long-term activity-dependent GABAergic synaptic plasticity.

Neuronal Cl^- Regulation

$GABA_AR$ activation allows the flux of Cl^- down its electrochemical gradient[2]. This flux depends upon a preexisting Cl^- gradient across the neuronal membrane which is largely set by the relative expression of the neuron-specific Cl^--extruding K^+-Cl^- cotransporter KCC2 (Payne et al. 1996; Rivera et al. 1999; Kahle et al. 2008; Blaesse et al. 2009) and the Cl^--intruding Na^+-K^+-$2Cl^-$ cotransporter NKCC1 (Yamada et al. 2004; Dzhala et al. 2005; Nakanishi et al. 2007). KCC2 and NKCC1 are both secondary active transporters that depend on ionic gradients established by the primary active transporter Na^+-K^+-ATPase. They are also both members of the cation-chloride cotransporter (CCC) gene family *SLC12a1-9* (Payne et al. 2003; Mercado et al. 2004; Gamba 2005).

During the embryonic period, NKCC1 is the prevalent CCC in immature neurons (Xu et al. 1994; Plotkin et al. 1997; Russell 2000; Dzhala et al. 2005) (Fig. 10.1a, c). NKCC1 uses energy from the inward Na^+ gradient to transport K^+ and Cl^- into the neuron (at a ratio of $1Na^+:1K^+:2Cl^-$). This elevated neuronal Cl^- concentration maintains E_{GABA} more depolarized than the action potential threshold. Thus, $GABA_AR$ activation results in Cl^- efflux, which depolarizes the membrane potential above action potential threshold, causing GABAergic excitation (Mueller et al. 1984; Ben-Ari et al. 1989; Luhmann and Prince 1991; Zhang et al. 1991). High levels of neuronal Cl^- have recently been shown to increase the decay time course of GABAergic currents (Houston et al. 2009). This suggests that during early development when neuronal Cl^- is high, $GABA_A$ receptor-mediated currents initiate more prolonged depolarization promoting neuronal excitability and voltage-gated calcium channel (VGCC) activation.

---

[1] Inhibitory synaptic plasticity can also be induced by pathological levels of neuronal activity such as those occurring during epileptic seizures, trauma, and pain. For a review see Kahle et al. 2008.

[2] The $GABA_AR$ is also permeable to HCO_3^- (which has an outward HCO_3^- gradient). As a result $E_{GABAA} > E_{Cl}$. For a review see Farrant and Kaila 2007.

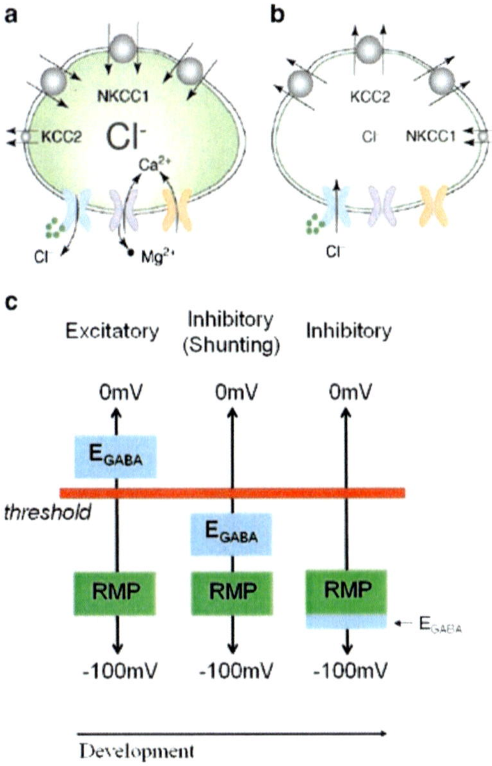

Fig. 10.1 Neuronal Cl^- regulation during development. (**a**) During embryonic development of the CNS intracellular Cl^- levels are relatively high due to the dominant expression of NKCC1. As a result, $GABA_AR$ activation (*blue*) leads to an efflux of Cl^- that depolarizes the membrane potential, which can permit Ca^{2+} influx via VGCCs (*orange*) and NMDARs (*purple*). (**b**) In the mature CNS the dominant expression of KCC2 renders intracellular Cl^- levels low, resulting in Cl^- influx upon $GABA_AR$ activation. (**c**) Schematic diagram illustrating the developmental transition of E_{GABA}. *Left*: E_{GABA} is depolarized with respect to the action potential threshold which renders GABAergic transmission excitatory. *Middle*: E_{GABA} is hyperpolarized to the action potential threshold but depolarized to the resting membrane potential (RMP). In this case GABAergic transmission is depolarizing but not excitatory. *Right*: E_{GABA} is hyperpolarized with respect to RMP, rendering GABAergic transmission inhibitory

Shortly after birth in rodents, NKCC1 is downregulated (Hubner et al. 2001; Yamada et al. 2004) and KCC2 is upregulated (Rivera et al. 1999)[3]. KCC2 derives energy from the K^+ gradient to extrude Cl^- against its gradient (at a ratio of $1K^+{:}1Cl^-$). As KCC2 extrudes Cl^-, E_{GABA} becomes hyperpolarized with respect to the resting membrane potential, which renders GABAergic transmission inhibitory (Rivera et al. 1999; Payne et al. 2003) (Fig. 10.1b, c).

[3]Not all species and brain regions exhibit the same timeline for the transition from excitation to inhibition. For a review see Blaesse et al. 2009.

Neuronal Activity and the Developmental Transition from GABAergic Excitation to Inhibition

Disrupting the normal expression pattern or regulation of NKCC1 or KCC2 during development alters neuronal Cl^-. This produces an imbalance of the inhibition-excitation ratio which is critical for proper neuronal circuit development and function (Turrigiano and Nelson 2004; Fukuda 2005; Hensch and Fagiolini 2005; Tao and Poo 2005; Akerman and Cline 2006; Kanold and Shatz 2006). Can activity itself function as a kind of feedback loop, affecting neuronal Cl^- regulation, and thus affecting the rapidity or efficacy of the developmental transition from excitatory to inhibitory GABA?

To examine the role of synaptic activity in the developmental transition from GABAergic excitation to inhibition Ganguly et al. blocked glutamatergic and GABAergic synaptic transmission pharmacologically[4]. While inhibiting glutamatergic transmission had no effect on the time course of the developmental transition, inhibition of depolarizing GABAergic transmission prevented the normal developmental hyperpolarization of E_{GABA} and reduced KCC2 upregulation (Ganguly et al. 2001). This suggested role of activity in the transition was supported by experiments where chronically depolarized cultured neurons showed enhanced KCC2 expression. However, blocking Na^+-dependent action potentials during development did not affect the time course of this transition, suggesting that spontaneous miniature GABAergic postsynaptic currents were sufficient to drive the transition. In this way, GABAergic transmission acts as a self-limiting factor in the developmental transition from depolarizing to hyperpolarizing action. These results are corroborated by in vivo experiments demonstrating that KCC2 upregulation during retinal development coincides with both the transition in GABA action and the abatement of propagating waves of activity (Vu et al. 2000; Sernagor et al. 2003). $GABA_AR$ inhibition with bicuculline during the period of the developmental transition resulted in GABA remaining excitatory, the waves of propagating activity continuing in the retinal ganglion cell layer, and the upregulation of KCC2 being inhibited (Leitch et al. 2005). Together these findings suggest a prominent role for GABAergic activity in the developmental transition from GABAergic excitation to inhibition. But this concept of GABA acting as a self-limiting factor is not without contention.

Ludwig et al. utilized a KCC2 antibody to examine the expression of KCC2 in mouse hippocampal cultures during development (Ludwig et al. 2003). They found a low level of KCC2 expression prior to the formation of the first synapses, indicating that synaptic transmission is not required for early KCC2 expression. Moreover, subsequent KCC2 upregulation was not hindered by preventing action potential

[4] $GABA_ARs$ were antagonized with bicuculline and picrotoxin; glutamatergic N-methyl-D-aspartate receptors (NMDARs) were blocked with D-APV, and non-NMDA receptors with CNQX.

firing, glutamatergic transmission, or GABAergic transmission (Ludwig et al. 2003), indicating a lack of requirement for $GABA_AR$ activation in the developmental transition. Experiments in cultured midbrain neurons show similar results; blocking $GABA_AR$ activity failed to alter the normal developmental transition (Titz et al. 2003). Moreover, an in vivo study demonstrated that NKCC1 disruption in neonatal mice which prevented excitatory $GABA_AR$-mediated responses failed to alter KCC2 upregulation (Sipila et al. 2009). Interestingly, a large increase in the intrinsic excitability of these neurons was noted, suggesting a compensatory mechanism that promotes homeostatic plasticity in the absence of normal depolarizing GABA activity (Sipila et al. 2009).

The presence of such a homeostatic drive offers a possible explanation for the discrepancy in results presented above. It is conceivable that such a drive is only present in some preparations or some areas of the brain, and thus the absence of $GABA_AR$ activation may result in different effects during different studies. Titz et al. does suggest that the fact that different brain regions were studied may be a factor in these discrepancies, and further notes that chronic $GABA_AR$ inhibition has been known to induce cell damage (Choi 1992), which can reduce KCC2 expression (van den Pol et al. 1996). Thus the nature and duration of the $GABA_AR$ inhibition in some studies could have resulted in damage to the neurons, which may have disguised actual developmental changes.

Synaptic Plasticity at Mature GABAergic Synapses

The majority of reported mechanisms underlying inhibitory GABAergic synaptic plasticity have been due to changes in conductance (Gaiarsa et al. 2002). However we now know that activity can regulate neuronal Cl^-, which is a powerful determinant of inhibitory transmission (Fiumelli and Woodin 2007). GABAergic synapses can be modified by pairing pre- and postsynaptic activity at low frequencies; this form of plasticity is termed GABAergic spike-timing dependent plasticity (STDP) (Woodin et al. 2003; Fiumelli and Woodin 2007; Balena and Woodin 2008; Saraga et al. 2008; Ormond and Woodin 2009b). STDP is a form of activity-induced plasticity that depends on the timing and temporal order of pre-and postsynaptic spiking (Dan and Poo 2006). At glutamatergic synapses in the mammalian CNS repeated coincident spiking of pre- and postsynaptic neurons (with a 15 ms delay) at a low frequency (1 Hz) induced LTP (Bi and Poo 1998). Reversing the order of spiking during plasticity induction (postsynaptic neuron before presynaptic neuron, versus pre before post), but maintaining the same time interval (−15 ms) produced long-term depression (LTD). This demonstration of STDP, along with those of other groups (Markram et al. 1997; Debanne et al. 1998), established the importance of spike timing in the induction of synaptic plasticity at glutamatergic synapses. But plasticity of inhibitory synapses has the potential to be just as important, given their ability to modulate network activity by shunting excitatory inputs (Miles 1990; Ormond and Woodin 2009b).

STDP of GABAergic synapses was first demonstrated at inhibitory GABAergic synapses in cultured hippocampal neurons (Woodin et al. 2003). Similar to excitatory synapses, the firing of coincident pre- and postsynaptic action potentials (at 5 Hz for 30 s) produces a long-term modification of inhibitory strength. Unlike glutamatergic synapses, where the spike-timing window is asymmetrical, GABAergic plasticity has a symmetrical spike-timing window. If the cells were fired within ±20 ms of each other, regardless of the pre/post order, then inhibition at the synapse was weakened (Woodin et al. 2003). Thus GABAergic STDP differs from glutamatergic STDP because it is independent of the temporal order of pre-and postsynaptic spiking, and instead depends of the correlation of spiking within a narrow time window.

The mechanism responsible for this weakening of inhibition is a shift in E_{GABA} to more depolarized values, resulting in less Cl^- influx upon $GABA_AR$ activation (Woodin et al. 2003). This depolarizing shift in E_{GABA} is due to a decrease in KCC2 activity, contingent upon Ca^{2+}-influx through voltage-gated calcium channels. Conversely, noncoincident spiking of pre- and postsynaptic neurons did not affect E_{GABA}, but rather weakened inhibition by decreasing $GABA_AR$ conductance (Woodin et al. 2003). This decrease may be due to a reduction in presynaptic transmitter release or a reduction in postsynaptic $GABA_A$ channel activity, but the precise mechanism is not currently known.

Activity-induced changes in Cl^- regulation of mature neurons do not always require presynaptic activity. Repetitive postsynaptic spiking alone also affects the strength of mature GABAergic synapses in rat hippocampal cultures. Spiking a neuron at a frequency of 10 Hz for 5 min (5,000 spikes total) led to a persistent depolarization of E_{GABA}, and thus a weakening of inhibition (Fiumelli et al. 2005). In this study E_{GABA} was determined by exogenously puffing GABA onto the spiked neuron. When the spiking frequency or the total duration of the spiking protocol were decreased, no plasticity resulted. An increase in intracellular Ca^{2+} concentration was required for this plasticity induction, with the magnitude of the increase corresponding to the magnitude of the plasticity. Blockage of L-type voltage-gated calcium channels prevented plasticity, as did the prevention of Ca^{2+} release from internal stores. However, it has also been shown that trains of action potentials, such as with the aforementioned postsynaptic spiking protocol, reset the thermodynamic equilibrium for NKCC1 transport and can indirectly lead to changes in neuronal Cl^- (Brumback and Staley 2008).

But what is the role of GABAergic STDP? How does it regulate network activity? These questions were recently addressed in the CA1 region of the hippocampus both experimentally (Ormond and Woodin 2009a, b) and computationally (Saraga et al. 2008). In this cortical circuit, feed-forward GABAergic inhibition is fast enough to shunt the peak of preceding excitatory postsynaptic potentials. When GABAergic STDP is induced at these feedforward inhibitory inputs, the resulting decrease in inhibition reduces the shunting of the preceding excitatory currents. This novel form of synaptic plasticity was termed disinhibition-mediated LTP (Ormond and Woodin 2009b). In effect it produces a long-term synapse-specific increase in the amplitude of Schaffer collateral-mediated postsynaptic potentials

(Ormond and Woodin 2009b). Given that disinhibition-mediated LTP shares the major features of classic glutamatergic LTP, it seems reasonable to speculate that it may be involved in the learning and memory functions of the hippocampus.

STDP and spiking-induced downregulation of KCC2 occurs within minutes of each other (Woodin et al. 2003; Fiumelli et al. 2005; Fiumelli and Woodin 2007), which would seem to preclude mechanisms such as changes in gene transcription or protein synthesis. Instead, alterations in membrane trafficking or posttranslational modifications to KCC2 seem more likely. Additionally, blocking Ca^{2+}-dependent protein kinase C (PKC) prevents the depolarization of E_{GABA} (Fiumelli et al. 2005); thus, it has been suggested that an increase in postsynaptic Ca^{2+} activates a PKC-dependent pathway that regulates KCC2 to produce synaptic plasticity (Fiumelli and Woodin 2007). No consensus has yet been reached as to the specific role of PKC; Fiumelli et al. concluded that PKC mediates the functional downregulation of KCC2 , whereas others state that PKC inhibits KCC2 endocytosis resulting in its upregulation (Lee et al. 2007).

Brain-derived neurotrophic factor (BDNF) can play a central role in activity-dependent Cl^--mediated GABAergic plasticity. BDNF increases KCC2 expression early in development (Aguado et al. 2003), and its absence, as well as that of its receptor TrkB, leads to decreased KCC2 expression in neonatal animals (Carmona et al. 2006). In mature animals, however, neuronal damage and hyperexcitation associated with pathophysiological conditions have been known to lower KCC2 expression through a BDNF/TrkB signaling pathway (Nabekura et al. 2002). Acute BDNF application to hippocampal cultures has also been shown to depolarize E_{GABA} through KCC2 downregulation (Wardle and Poo 2003)[5].

Synaptic Plasticity at Immature GABAergic Synapses

Inhibitory synapses are sensitive to spike timing in the mature CNS. As explained above, paired activity modifies inhibitory synapses through a regulation of KCC2. But how would synapses respond early in development, before KCC2 is upregulated? Coincident pre-and postsynaptic activity at depolarizing GABAergic synapses[6] led to modifications in synaptic strength in both hippocampal cultures (Balena and Woodin 2008) and slices (Xu et al. 2008). This coincident spiking at 5 Hz hyperpolarized E_{GABA}, which strengthened inhibition (Balena and Woodin 2008; Xu et al. 2008); this is the reverse of what was seen at mature synapses. Application of the NKCC1 antagonist bumetanide hyperpolarized E_{GABA} as previously demonstrated (Staley and Mody 1992), and prevented any subsequent activity-induced shifts in E_{GABA} (Balena

[5] Further information on the role of neurotrophic factors at inhibitory synapses can be found in the Chapter by Gaiarsa et al. in this volume.

[6] At synapses examined in these studies E_{GABA} lay between the resting membrane potential and the action potential threshold; thus, these synapses were depolarizing, but still inhibitory.

and Woodin 2008; Xu et al. 2008). However, increasing the stimulation frequency to 20 Hz or greater resulted in E_{GABA} depolarization (Xu et al. 2008). Regardless of the stimulation frequency, all plasticity induction was dependent on NKCC1 and Ca^{2+}-influx via L-type VGCCs. Higher stimulation frequencies (≥20 Hz) also required the activity of $GABA_BRs$ and CaMKII (Xu et al. 2008). Xu et al. theorized that GABA "spillover" occurred at higher spiking frequencies, and that this was sufficient to recruit $GABA_BRs$ and transition the direction of the E_{GABA} shift from hyperpolarization to depolarization. This depolarizing shift could allow increased time for dendritic arborization and synaptogenesis (Xu et al. 2008).

Conclusion and Future Directions

It is now clear that physiological levels of activity can modify inhibitory synapses via regulation of neuronal Cl^-. A potential reason for the relative late-coming of the elucidation of neuronal Cl^- regulation as a mechanism underlying inhibitory synaptic plasticity, compared to that of the mechanisms underlying glutamatergic LTP, is the technical difficulty associated with maintaining physiological Cl^- gradients during electrophysiological recordings. The first report of gramicidin perforated patch clamp recordings, which prevent artifactual changes in intracellular Cl^-, was in 1995 (Kyrozis and Reichling 1995); this was followed in 2003 with a non-invasive method for recording E_{GABA} using cell-attached recordings (Tyzio et al. 2003). It wasn't until the advent of these electrophysiological advances that activity-induced changes neuronal Cl^- could be readily resolved.

What remains to be elucidated, at the cellular level, are the precise mechanisms involved in plasticity induction, as well as the compartment-specific regulation of Cl^-. These efforts have been hindered by a lack of specific antagonists for KCC2 (Payne et al. 2003). However a recent small molecule screen identified inhibitors of KCC2 (Delpire et al. 2009); hopefully, further characterization of these molecules will confirm them as suitable pharmacological agents for the specific antagonism of KCC2.

Emphasis also needs to be placed on understanding how simultaneous inhibitory and excitatory synaptic plasticity impact network activity throughout the brain. This will serve to shed new light on broader processes that plasticity is thought to underlie, such as perception and memory.

References

Aguado F, Carmona MA, Pozas E, Aguilo A, Martinez-Guijarro FJ, Alcantara S, Borrell V, Yuste R, Ibanez CF, Soriano E (2003) BDNF regulates spontaneous correlated activity at early developmental stages by increasing synaptogenesis and expression of the K^+/Cl^- co-transporter KCC2. Development 130:1267–1280.

Aizenman CD, Manis PB, Linden DJ (1998) Polarity of long-term synaptic gain change is related to postsynaptic spike firing at a cerebellar inhibitory synapse. Neuron 21:827–835.
Akerman CJ, Cline HT (2006) Depolarizing GABAergic conductances regulate the balance of excitation to inhibition in the developing retinotectal circuit in vivo. J Neurosci 26:5117–5130.
Balena T, Woodin MA (2008) Coincident pre- and postsynaptic activity downregulates NKCC1 to hyperpolarize E(Cl) during development. Eur J Neurosci 27:2402–2412.
Ben-Ari Y, Cherubini E, Corradetti R, Gaiarsa JL (1989) Giant synaptic potentials in immature rat CA3 hippocampal neurones. J Physiol 416:303–325.
Ben-Ari Y, Gaiarsa JL, Tyzio R, Khazipov R (2007) GABA: a pioneer transmitter that excites immature neurons and generates primitive oscillations. Physiol Rev 87:1215–1284.
Berlau DJ, McGaugh JL (2006) Enhancement of extinction memory consolidation: the role of the noradrenergic and GABAergic systems within the basolateral amygdala. Neurobiol Learn Mem 86:123–132.
Bi GQ, Poo MM (1998) Synaptic modifications in cultured hippocampal neurons: dependence on spike timing, synaptic strength, and postsynaptic cell type. J Neurosci 18:10464–10472.
Blaesse P, Airaksinen MS, Rivera C, Kaila K (2009) Cation-chloride cotransporters and neuronal function. Neuron 61:820–838.
Bliss TV, Lomo T (1973) Long-lasting potentiation of synaptic transmission in the dentate area of the anaesthetized rabbit following stimulation of the perforant path. J Physiol 232: 331–356.
Bliss TV, Collingridge GL (1993) A synaptic model of memory: long-term potentiation in the hippocampus. Nature 361:31–39.
Brumback AC, Staley KJ (2008) Thermodynamic regulation of NKCC1-mediated Cl^- cotransport underlies plasticity of GABA(A) signaling in neonatal neurons. J Neurosci 28:1301–1312.
Caillard O, Ben-Ari Y, Gaiarsa JL (1999) Mechanisms of induction and expression of long-term depression at GABAergic synapses in the neonatal rat hippocampus. J Neurosci 19:7568–7577.
Caillard O, Ben-Ari Y, Gaiarsa JL (2000) Activation of presynaptic and postsynaptic ryanodine-sensitive calcium stores is required for the induction of long-term depression at GABAergic synapses in the neonatal rat hippocampus. J Neurosci 20:RC94.
Carmona MA, Pozas E, Martinez A, Espinosa-Parrilla JF, Soriano E, Aguado F (2006) Age-dependent spontaneous hyperexcitability and impairment of GABAergic function in the hippocampus of mice lacking trkB. Cereb Cortex 16:47–63.
Choi DW (1992) Excitotoxic cell death. J Neurobiol 23:1261–1276.
Dan Y, Poo MM (2006) Spike timing-dependent plasticity: from synapse to perception. Physiol Rev 86:1033–1048.
Debanne D, Gahwiler BH, Thompson SM (1998) Long-term synaptic plasticity between pairs of individual CA3 pyramidal cells in rat hippocampal slice cultures. J Physiol 507 (Pt 1):237–247.
Delpire E, Days E, Lewis LM, Mi D, Kim K, Lindsley CW, Weaver CD (2009) Small-molecule screen identifies inhibitors of the neuronal K-Cl cotransporter KCC2. Proc Natl Acad Sci USA 106:5383–5388.
Dzhala VI, Talos DM, Sdrulla DA, Brumback AC, Mathews GC, Benke TA, Delpire E, Jensen FE, Staley KJ (2005) NKCC1 transporter facilitates seizures in the developing brain. Nat Med 11:1205–1213.
Farrant M, Kaila K (2007) The cellular, molecular and ionic basis of GABA(A) receptor signalling. Prog Brain Res 160:59–87.
Fiumelli H, Woodin MA (2007) Role of activity-dependent regulation of neuronal chloride homeostasis in development. Curr Opin Neurobiol 17:81–86.
Fiumelli H, Cancedda L, Poo MM (2005) Modulation of GABAergic transmission by activity via postsynaptic Ca2+-dependent regulation of KCC2 function. Neuron 48:773–786.
Fukuda A (2005) Diuretic soothes seizures in newborns. Nat Med 11:1153–1154.

Gaiarsa JL, Caillard O, Ben-Ari Y (2002) Long-term plasticity at GABAergic and glycinergic synapses: mechanisms and functional significance. Trends Neurosci 25:564–570.

Gamba G (2005) Molecular physiology and pathophysiology of electroneutral cation-chloride cotransporters. Physiol Rev 85:423–493.

Ganguly K, Schinder AF, Wong ST, Poo M (2001) GABA itself promotes the developmental switch of neuronal GABAergic responses from excitation to inhibition. Cell 105:521–532.

Hensch TK, Fagiolini M (2005) Excitatory-inhibitory balance and critical period plasticity in developing visual cortex. Prog Brain Res 147:115–124.

Houston CM, Bright DP, Sivilotti LG, Beato M, Smart TG (2009) Intracellular chloride ions regulate the time course of GABA-mediated inhibitory synaptic transmission. J Neurosci 29:10416–10423.

Hubner CA, Stein V, Hermans-Borgmeyer I, Meyer T, Ballanyi K, Jentsch TJ (2001) Disruption of KCC2 reveals an essential role of K-Cl cotransport already in early synaptic inhibition. Neuron 30:515–524.

Kahle KT, Staley KJ, Nahed BV, Gamba G, Hebert SC, Lifton RP, Mount DB (2008) Roles of the cation-chloride cotransporters in neurological disease. Nat Clin Pract Neurol 4:490–503.

Kaila K (1994) Ionic basis of GABAA receptor channel function in the nervous system. Prog Neurobiol 42:489–537.

Kanold PO, Shatz CJ (2006) Subplate neurons regulate maturation of cortical inhibition and outcome of ocular dominance plasticity. Neuron 51:627–638.

Katz LC, Shatz CJ (1996) Synaptic activity and the construction of cortical circuits. Science 274:1133–1138.

Komatsu Y, Iwakiri M (1993) Long-term modification of inhibitory synaptic transmission in developing visual cortex. Neuroreport 4:907–910.

Kyrozis A, Reichling DB (1995) Perforated-patch recording with gramicidin avoids artifactual changes in intracellular chloride concentration. J Neurosci Methods 57:27–35.

Lee HH, Walker JA, Williams JR, Goodier RJ, Payne JA, Moss SJ (2007) Direct protein kinase C-dependent phosphorylation regulates the cell surface stability and activity of the potassium chloride cotransporter KCC2. J Biol Chem 282:29777–29784.

Lee WC, Huang H, Feng G, Sanes JR, Brown EN, So PT, Nedivi E (2006) Dynamic remodeling of dendritic arbors in GABAergic interneurons of adult visual cortex. PLoS Biol 4:e29.

Leitch E, Coaker J, Young C, Mehta V, Sernagor E (2005) GABA type-A activity controls its own developmental polarity switch in the maturing retina. J Neurosci 25:4801–4805.

Lien CC, Mu Y, Vargas-Caballero M, Poo MM (2006) Visual stimuli-induced LTD of GABAergic synapses mediated by presynaptic NMDA receptors. Nat Neurosci 9:372–380.

Ludwig A, Li H, Saarma M, Kaila K, Rivera C (2003) Developmental up-regulation of KCC2 in the absence of GABAergic and glutamatergic transmission. Eur J Neurosci 18:3199–3206.

Luhmann HJ, Prince DA (1991) Postnatal maturation of the GABAergic system in rat neocortex. J Neurophysiol 65:247–263.

Malenka RC, Bear MF (2004) LTP and LTD: an embarrassment of riches. Neuron 44:5–21.

Markram H, Lubke J, Frotscher M, Sakmann B (1997) Regulation of synaptic efficacy by coincidence of postsynaptic APs and EPSPs. Science 275:213–215.

Martin SJ, Grimwood PD, Morris RG (2000) Synaptic plasticity and memory: an evaluation of the hypothesis. Annu Rev Neurosci 23:649–711.

McLean HA, Caillard O, Ben-Ari Y, Gaiarsa JL (1996) Bidirectional plasticity expressed by GABAergic synapses in the neonatal rat hippocampus. J Physiol 496 (Pt 2):471–477.

Mercado A, Mount DB, Gamba G (2004) Electroneutral cation-chloride cotransporters in the central nervous system. Neurochem Res 29:17–25.

Miles R (1990) Synaptic excitation of inhibitory cells by single CA3 hippocampal pyramidal cells of the guinea-pig in vitro. J Physiol 428:61–77.

Mueller AL, Taube JS, Schwartzkroin PA (1984) Development of hyperpolarizing inhibitory postsynaptic potentials and hyperpolarizing response to gamma-aminobutyric acid in rabbit hippocampus studied in vitro. J Neurosci 4:860–867.

Nabekura J, Ueno T, Okabe A, Furuta A, Iwaki T, Shimizu-Okabe C, Fukuda A, Akaike N (2002) Reduction of KCC2 expression and GABAA receptor-mediated excitation after in vivo axonal injury. J Neurosci 22:4412–4417.
Nakanishi K, Yamada J, Takayama C, Oohira A, Fukuda A (2007) NKCC1 activity modulates formation of functional inhibitory synapses in cultured neocortical neurons. Synapse 61:138–149.
Ormond J, Woodin MA (2009a) A heterosynaptic increase in GABAergic conductance maintains the pathway specificity of disinhibition-mediated LTP. Journal of Neuroscience Submitted.
Ormond J, Woodin MA (2009b) Disinhibition mediates a form of hippocampal long-term potentiation in area CA1. PLoS One 4:e7224.
Payne JA, Stevenson TJ, Donaldson LF (1996) Molecular characterization of a putative K-Cl cotransporter in rat brain. A neuronal-specific isoform. J Biol Chem 271:16245–16252.
Payne JA, Rivera C, Voipio J, Kaila K (2003) Cation-chloride co-transporters in neuronal communication, development and trauma. Trends Neurosci 26:199–206.
Plotkin MD, Snyder EY, Hebert SC, Delpire E (1997) Expression of the Na-K-2Cl cotransporter is developmentally regulated in postnatal rat brains: a possible mechanism underlying GABA's excitatory role in immature brain. J Neurobiol 33:781–795.
Ramirez DR, Buzzetti RA, Savage LM (2005) The role of the GABA(A) agonist muscimol on memory performance: reward contingencies determine the nature of the deficit. Neurobiol Learn Mem 84:184–191.
Rivera C, Voipio J, Payne JA, Ruusuvuori E, Lahtinen H, Lamsa K, Pirvola U, Saarma M, Kaila K (1999) The K^+/Cl^- co-transporter KCC2 renders GABA hyperpolarizing during neuronal maturation. Nature 397:251–255.
Russell JM (2000) Sodium-potassium-chloride cotransport. Physiol Rev 80:211–276.
Saraga F, Balena T, Wolansky T, Dickson CT, Woodin MA (2008) Inhibitory synaptic plasticity regulates pyramidal neuron spiking in the rodent hippocampus. Neuroscience 155:64–75.
Sernagor E, Young C, Eglen SJ (2003) Developmental modulation of retinal wave dynamics: shedding light on the GABA saga. J Neurosci 23:7621–7629.
Sipila ST, Huttu K, Yamada J, Afzalov R, Voipio J, Blaesse P, Kaila K (2009) Compensatory enhancement of intrinsic spiking upon NKCC1 disruption in neonatal hippocampus. J Neurosci 29:6982–6988.
Staley KJ, Mody I (1992) Shunting of excitatory input to dentate gyrus granule cells by a depolarizing GABAA receptor-mediated postsynaptic conductance. J Neurophysiol 68:197–212.
Tao HW, Poo MM (2005) Activity-dependent matching of excitatory and inhibitory inputs during refinement of visual receptive fields. Neuron 45:829–836.
Titz S, Hans M, Kelsch W, Lewen A, Swandulla D, Misgeld U (2003) Hyperpolarizing inhibition develops without trophic support by GABA in cultured rat midbrain neurons. J Physiol 550:719–730.
Turrigiano GG, Nelson SB (2004) Homeostatic plasticity in the developing nervous system. Nat Rev Neurosci 5:97–107.
Tyzio R, Ivanov A, Bernard C, Holmes GL, Ben-Ari Y, Khazipov R (2003) Membrane potential of CA3 hippocampal pyramidal cells during postnatal development. J Neurophysiol 90:2964–2972.
van den Pol AN, Obrietan K, Chen G (1996) Excitatory actions of GABA after neuronal trauma. J Neurosci 16:4283–4292.
Vu TQ, Payne JA, Copenhagen DR (2000) Localization and developmental expression patterns of the neuronal K-Cl cotransporter (KCC2) in the rat retina. J Neurosci 20:1414–1423.
Wardle RA, Poo MM (2003) Brain-derived neurotrophic factor modulation of GABAergic synapses by postsynaptic regulation of chloride transport. J Neurosci 23:8722–8732.
Woodin MA, Ganguly K, Poo MM (2003) Coincident pre- and postsynaptic activity modifies GABAergic synapses by postsynaptic changes in Cl^- transporter activity. Neuron 39:807–820.

Xu C, Zhao MX, Poo MM, Zhang XH (2008) GABA(B) receptor activation mediates frequency-dependent plasticity of developing GABAergic synapses. Nat Neurosci 11:1410–1418.

Xu JC, Lytle C, Zhu TT, Payne JA, Benz E, Jr., Forbush B, 3rd (1994) Molecular cloning and functional expression of the bumetanide-sensitive Na-K-Cl cotransporter. Proc Natl Acad Sci USA 91:2201–2205.

Yamada J, Okabe A, Toyoda H, Kilb W, Luhmann HJ, Fukuda A (2004) Cl^- uptake promoting depolarizing GABA actions in immature rat neocortical neurones is mediated by NKCC1. J Physiol 557:829–841.

Zhang L, Spigelman I, Carlen PL (1991) Development of GABA-mediated, chloride-dependent inhibition in CA1 pyramidal neurones of immature rat hippocampal slices. J Physiol 444:25–49.

Chapter 11
Endocannabinoid Mediated Long-Term Depression at Inhibitory Synapses

Chiayu Q. Chiu and Pablo E. Castillo

Abstract Long-lasting activity-dependent changes in synaptic strength, in the form of long-term potentiation (LTP) or depression (LTD), are thought to be the cellular basis of learning and memory. Although inhibitory synapses are critical for proper functioning of neural circuits, most of the studies investigating synaptic plasticity have concentrated on excitatory glutamatergic synapses. The idea that inhibitory synapses are plastic like excitatory synapses is gradually being accepted. Different forms of GABAergic synaptic plasticity have recently been reported. Among them, one of the most intensely studied is a form of LTD mediated by a group of retrograde messengers, collectively called endocannabinoids (eCBs). eCB-mediated LTD at inhibitory synapses (I-LTD) is a heterosynaptic form of plasticity, whose induction typically requires the activation of metabotropic glutamatergic receptors by nearby excitatory inputs and the resulting eCB mobilization from the postsynaptic cell. By activating presynaptic type 1 cannabinoid receptors (CB1Rs), eCBs can cause a long-lasting reduction in GABA release. I-LTD has been identified in several brain structures and may serve as a mechanism by which neurons adjust the strength of the inhibition they receive in response to excitatory afferent stimulation. In light of the wide-ranging effects of inhibition on synaptic transmission, from shaping the input-output relationship and excitability of neurons to modulating the inducibility of excitatory synaptic plasticity, I-LTD is expected to have significant impact on the excitatory/inhibitory balance within circuits and on neural network function. In this chapter, we review the main properties of eCB-mediated LTD at GABAergic synapses with the aim of understanding its physiological role.

P.E. Castillo (✉)
Dominick P. Purpura Department of Neuroscience, Albert Einstein College of Medicine, Bronx, NY 10461, USA
e-mail: pablo.castillo@einstein.yu.edu

M.A. Woodin and A. Maffei (eds.), *Inhibitory Synaptic Plasticity*,
DOI 10.1007/978-1-4419-6978-1_11,

Introduction

The eCBs are lipid molecules that get their name from the more well-known exogenous cannabinoid counterpart, Δ^9-tetrahydrocannabinol (Δ^9-THC), the main psychoactive component in marijuana. Long before the synaptic effects of cannabinoids or eCBs were elucidated in the brain, it was known anecdotally that smoking marijuana disrupts learning and memory in humans. These disruptive effects by cannabinoids were later confirmed in controlled laboratory settings using human and animal subjects (for reviews on this topic, see Hampson and Deadwyler 1998; Sullivan 2000; Ranganathan and D'Souza 2006). For example, individuals under the influence of marijuana show impaired short-term memory, performing worse in behavioral tasks like word recall. Likewise, administering Δ^9-THC and other cannabinoidmimetic agents to laboratory animals disrupt their performance in a variety of memory tasks, including delayed nonmatch-to-sample for nonhuman primates (Zimmerberg et al. 1971; Aigner 1988) and the hidden Morris water maze for rodents (Lichtman et al. 1995; Ferrari et al. 1999; Varvel et al. 2001). Notably, laboratory animals whose eCB system is disrupted by pharmacological or genetic manipulations show deficits in fear extinction (Marsicano et al. 2002; Suzuki et al. 2004) and working memory (Varvel and Lichtman 2002). However, given the pervasive presence of the eCB system throughout the central nervous system (CNS) (Freund et al. 2003; Chevaleyre et al. 2006; Heifets and Castillo 2009; Kano et al. 2009), the neural mechanisms underlying these behavioral effects are not well understood.

Endocannabinoid Signaling and Synaptic Function

In order to gain a better understanding of the complexity underlying eCB signaling, we will briefly review the components of the eCB system, which comprises the eCBs, their synthesizing and metabolizing enzymes, putative transporters and the cannabinoid receptors. The eCBs are a group of structurally distinct lipophilic molecules, each with their own machinery for synthesis and degradation (Piomelli 2003; Ligresti et al. 2005), that act as ligands for the two cloned cannabinoid receptors. The type 1 cannabinoid receptor (CB1R) is expressed predominantly in the CNS (Matsuda et al. 1990) and the type 2 cannabinoid receptor (CB2R) is mostly present peripherally in the immune system (Munro et al. 1993). Both are seven transmembrane spanning receptors that are coupled to the pertussis toxin sensitive G protein, $G_{i/o}$ (Howlett 2002). Consequently, activation of either cannabinoid receptor reduces adenylyl cyclase (AC) activity, thus lowering cAMP levels and protein kinase A (PKA) activity. Due to the focus of this chapter on CNS synapses, CB2Rs will not be discussed further. Interestingly, CB1R is one of the most highly expressed G protein-coupled receptors in the brain (Herkenham et al. 1990), suggesting that the eCB system is involved in many facets of CNS function. At the subcellular level, CB1Rs are typically found at axon terminals

(Egertova et al. 1998; Katona et al. 1999; Bodor et al. 2005), consistent with its proposed role in controlling neurotransmitter release. In addition to the decrease in cAMP-PKA signaling, inhibition of voltage gated calcium channels (VGCCs) and activation of potassium channels at the presynaptic site are likely mechanisms underlying eCB-mediated suppression of transmitter release in the CNS (Howlett et al. 2004; Lovinger 2008; Kano et al. 2009).

Several eCBs have been identified from the mammalian brain (Sugiura et al. 2002; Kano et al. 2009). The first one to be isolated was *N*-arachidonoylethanolamide, otherwise known as anandamide (AEA) from the Sanskrit word meaning *bliss* (Devane et al. 1992). Intriguingly, AEA behaves as a nonselective partial agonist for CB1Rs and CB2Rs (Zygmunt et al. 1999; Smart et al. 2000; Sugiura et al. 2002). Subsequently, 2-arachidonoylglycerol (2-AG) was isolated (Mechoulam et al. 1995; Sugiura et al. 1995). 2-AG is considered a major eCB given that it is found at high concentrations in the brain and acts as a specific full cannabinoid receptor agonist. Other putative eCBs include dihomo-γ-linolenoyl ethanolamide, docosatetraenoyl ethanolamide, 2-arachidonoyl glycerol ether (noladin ether), *O*-arachidonoylethanolamide (virodhamine) and *N*-arachidonoyldopamine (NADA). The roles of these other eCBs in synaptic transmission are not yet clear.

The first step in eCB-mediated synaptic plasticity is the activity-dependent synthesis of eCBs in the postsynaptic cell. Typically, eCBs are produced following either membrane depolarization (by way of intracellular calcium) or bursts of synaptic activity (by way of metabotropic receptor activation). Interestingly, eCB production is facilitated under conditions in which both calcium and metabotropic pathways are simultaneously activated (Hashimotodani et al. 2007), as would occur when afferent activity coincides with postsynaptic firing. Once synthesized, eCBs are mobilized from the postsynaptic plasma membrane, possibly via facilitated transport by an unidentified eCB membrane transporter (EMT), which may function bidirectionally in eCB reuptake from synapses (Beltramo et al. 1997; Bisogno et al. 2001). After retrogradely crossing the synaptic cleft, eCBs activate CB1Rs that are present at both glutamatergic and GABAergic terminals, thereby reducing transmitter release at both excitatory and inhibitory synapses, in a transient or consolidated manner (Chevaleyre et al. 2006). Short-term depression (STD) is commonly triggered by depolarization of the postsynaptic cell, leading to a short-lived calcium-dependent mobilization of eCBs and transient CB1R activation. This process is called depolarization-induced suppression of inhibition (DSI) or excitation (DSE) depending on the inhibitory or excitatory nature of the target synapse (Alger 2002; Kano et al. 2009). eCB-mediated STD can also be evoked with short bursts of synaptic activity (Maejima et al. 2001; Brown et al. 2003; Galante and Diana 2004; Best and Regehr 2008). Brief CB1R activation causes STD by primarily inhibiting presynaptic VGCCs (Wilson et al. 2001; Brown et al. 2003). On the other hand, eCB-mediated LTD is typically initiated following repetitive glutamatergic synaptic activity and may involve modulation of the presynaptic vesicular release machinery (Heifets and Castillo 2009). How activation of CB1Rs leads to STD or LTD, as well as the precise mechanism by which CB1R activation triggers long-lasting reduction of transmitter release are active areas of investigation.

Endocannabinoid-Mediated LTD of GABAergic Synaptic Transmission

Synaptically induced eCB-LTD has been described in several brain regions at both excitatory synapses (E-LTD) and inhibitory synapses (I-LTD) (Heifets and Castillo 2009). E-LTD has been observed in dorsal striatum (Gerdeman et al. 2002; Kreitzer and Malenka 2005), nucleus accumbens (Robbe et al. 2002), visual cortex (Sjostrom et al. 2003; Crozier et al. 2007), somatosensory cortex (Bender et al. 2006; Nevian and Sakmann 2006), prefrontal cortex (Lafourcade et al. 2007), neonatal hippocampus (Yasuda et al. 2008), cerebellum (Safo and Regehr 2005; Soler-Llavina and Sabatini 2006), and brain stem (Tzounopoulos et al. 2007; Penzo and Pena 2009), while I-LTD has been reported in basolateral amygdala (Marsicano et al. 2002), hippocampus (Chevaleyre and Castillo 2003), VTA (Pan et al. 2008a), dorsal striatum (Adermark and Lovinger 2009), prefrontal cortex (Chiu et al. 2010), and also in cortico-tectal co-cultures (Henneberger et al. 2007). It is therefore clear that eCB-mediated LTD is a widespread phenomenon in the brain. These studies have also revealed three important mechanistic properties of eCB-mediated LTD: first, induction of E-LTD and I-LTD is abolished by pharmacological blockade of CB1Rs and is absent in transgenic mice lacking CB1Rs; second, eCB mobilization required for induction typically occurs as a result of group I metabotropic glutamate receptors (mGluR-I) and third, expression is associated with a long-lasting reduction in the probability of transmitter release. As for inhibitory synapses, given the mGluR-I requirement for eCB mobilization, it follows that I-LTD is a heterosynaptic form of plasticity (Fig. 11.1a).

The functional impact of eCB-mediated E-LTD or I-LTD on the propagation of signals and neural activity within networks is remarkably different. While depression of excitatory transmission would decrease propagation of activity, depression of inhibitory transmission is expected to enhance excitability and promote information transfer. Moreover, by removing inhibition in a long-lasting manner, the induction of I-LTD facilitates subsequent induction of "classical" LTP at excitatory synapses (E-LTP). In the following sections, we will address specific properties of eCB-mediated I-LTD reported in various central inhibitory synapses.

Fig. 11.1 (continued) (*right*) reported in the amygdala (Azad et al. 2004) may underlie AEA formation. Similar pathway could be involved in striatal I-LTD. The adenylyl cyclase (AC)-protein kinase A (PKA) cascade is linked to AEA synthesis via an unknown mechanism. In striatum, calcium influx through L-type calcium channels may facilitate this process. Once produced, mobilization of 2-AG and AEA may be facilitated by a putative eCB membrane transporter (EMT). (**c**) Expression of I-LTD as reported in the hippocampus. Activation of CB1Rs inhibits presynaptic PKA activity. Consolidation of I-LTD requires activity of inhibitory interneurons during induction to elevate calcium, which then stimulates the calcium-sensitive phosphatase, calcineurin (CaN). Coincident CaN activation and PKA inhibition shifts the phosphorylation status of a yet unidentified substrate (S), which may directly underlie the long-term reduction in vesicular GABA release

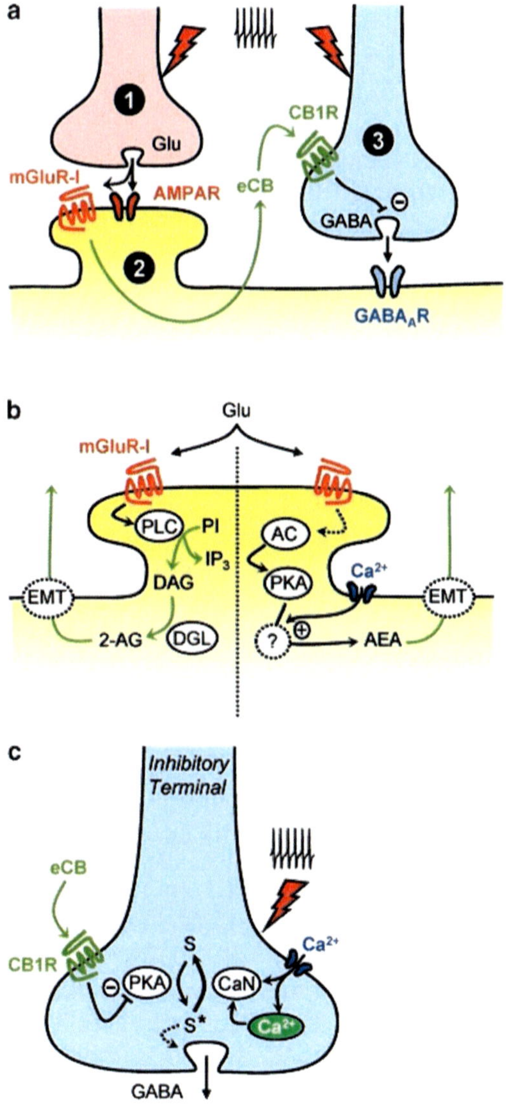

Fig. 11.1 Molecular mechanisms underlying heterosynaptic I-LTD. (**a**) Simplified diagram depicts the sequence of events leading to I-LTD. Induction of I-LTD is initiated by bursts of excitatory afferent activity (1) which releases glutamate to activate group I metabotropic glutamate receptors (mGluR-I). Activation of postsynaptic mGluR-I stimulates production of eCBs (2) that then cross the synaptic cleft and retrogradely bind to presynaptic CB1Rs on nearby inhibitory terminals. Finally, activation of CB1Rs in conjunction with interneuronal activity leads to a long-term reduction in GABA release (3). (**b**) Two signaling pathways downstream of mGluR-I can mediate eCB mobilization required for the induction of I-LTD. The first one (*left*), originally described in the hippocampus (Chevaleyre and Castillo 2003), involves the hydrolysis of phosphatidylinositol (PI) by phospholipase C (PLC) into diacylglycerol (DAG) and inositol 1,4,5-triphosphate (IP3). The DAG product is then converted into the eCB 2-AG by diacylglycerol lipase (DGL). An alternative pathway

Basolateral Amygdala

The role of eCBs in LTD at GABAergic synapses was first reported in the amygdala (Marsicano et al. 2002), a critical region for aversive learning (Maren and Quirk 2004; Rodrigues et al. 2004). Indeed, low frequency stimulation (LFS; 100 pulses at 1 Hz) of synaptic inputs triggered LTD of pharmacologically isolated inhibitory synaptic responses (IPSCs) in principal neurons of the basolateral amygdala, which the authors termed LTDi (Marsicano et al. 2002). To be consistent with the terminology used thus far in the chapter, we will refer to LTDi as I-LTD. This LFS-induced I-LTD was associated with a change in paired-pulse ratio (PPR), indicative of a presynaptic reduction in GABA release. The presynaptic locus of I-LTD expression was later supported by observations that spontaneous IPSC frequency but not amplitude was reduced following I-LTD (Azad et al. 2004). In addition, mGluR-I activation, specifically mGluR1 is not only necessary but also sufficient to trigger this form of plasticity (Azad et al. 2004). Induction of I-LTD most likely occurs in the postsynaptic cell because it is abolished by blocking intracellular G-protein signaling with GTPγS. However, unlike DSI, I-LTD does not require intracellular calcium. Two observations implicate AEA in amygdalar I-LTD: first, the magnitude of I-LTD is larger in mice that lack FAAH, the enzyme presumed to be responsible for metabolic removal of AEA, and second, inhibition of phospholipase C (PLC) and diacylglycerol lipase (DGL), the main enzymes that mediate 2-AG production, does not affect I-LTD. Currently, it is unknown how mGluR-I activation leads to AEA production. Curiously, loading postsynaptic cells with inhibitors of AC or PKA also abolished I-LTD, suggesting that postsynaptic AC-PKA signaling cascade may also play a role in eCB mobilization (Azad et al. 2004) (Fig. 11.1b). Future experiments will be needed to determine whether AC-PKA signaling simply permits I-LTD induction or actively participates in the induction process.

Increasing evidence indicates that local inhibitory circuits in the amygdala play an important role in fear learning (Ehrlich et al. 2009), raising the possibility that eCBs, via amygdalar I-LTD, may participate in this process. Consistent with this idea, animals treated with the CB1R antagonist SR141716A as well as CB1R-knockout mice show an impaired ability to extinguish a previously learned fear association (Marsicano et al. 2002). Extinction of aversive memories is partially based on an active relearning of a new association (Myers and Davis 2007), which may be encoded as LTP at excitatory synapses in the amygdala. eCB-mediated I-LTD promotes subsequent LTP induction at excitatory synapses in the amygdala and enhances basolateral amygdalar output to the central nucleus (Azad et al. 2004). Thus, I-LTD in the amygdala may participate in fear extinction by facilitating encoding of the extinction memory trace at the cellular level either directly by regulating inhibitory transmission or indirectly by modulating excitatory plasticity. The specific contribution of I-LTD in amygdalar-dependent learning remains to be determined.

Hippocampus

In the hippocampus, a crucial brain structure for learning and memory (Milner and Penfield 1955; Barnes 1988; O'Keefe 1999), repetitive stimulation of CA3 excitatory afferents to CA1 pyramidal neurons induces eCB-mediated LTD of nearby GABAergic inputs (Chevaleyre and Castillo 2003, 2004), a phenomenon independently confirmed by several groups (Edwards et al. 2006; Kang-Park et al. 2007; Zhu and Lovinger 2007). One study demonstrated hippocampal I-LTD in adolescent but not adult rats (Kang-Park et al. 2007), suggesting that eCB signaling may be developmentally down-regulated. Importantly, while CB1R activation is necessary for the induction of hippocampal I-LTD, the expression of this form of plasticity becomes CB1R-independent (Fig. 11.2a).

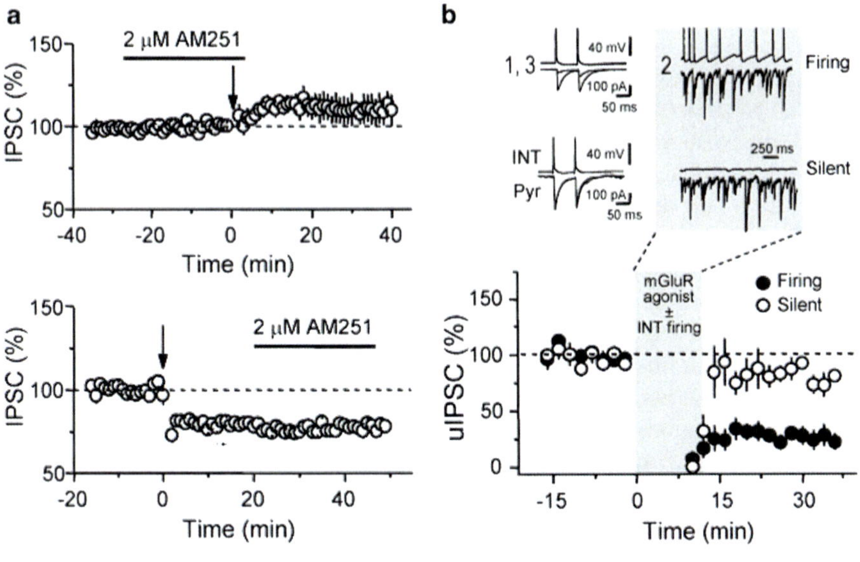

Fig. 11.2 Dependence of I-LTD on CB1R activation and interneuronal activity. (**a**) Activation of CB1Rs is required for I-LTD induction but not maintenance. In the hippocampus, delivery of tetanus (*vertical arrow*) typically induced a long-lasting reduction in evoked inhibitory responses (IPSCs) (*bottom*). When the CB1R antagonist AM251 was present during tetanus delivery, I-LTD was blocked (*top*). In contrast, when bath applied 10 minutes after delivering the tetanus, AM251 exhibited no effect on I-LTD (*bottom*). [Reproduced with permission from Chevaleyre and Castillo 2003 (Neuron vol 38, p 461)]. (**b**) Activity of CB1R-positive GABAergic interneurons is required for I-LTD as demonstrated in cell pair recordings at interneuron (INT) to pyramidal cell (Pyr) synapses in the hippocampus. Activation of mGluR-I triggers I-LTD (*bottom, black circles*) when the presynaptic interneuron is firing during agonist application (*upper inset, gray area*). However, if interneuron firing was suppressed by hyperpolarization (*Silent*), I-LTD was abolished (*bottom, white circles*). Average traces of unitary IPSCs (uIPSC), with the evoking presynaptic action potentials, are shown for the firing and silent condition (*upper inset, left*). [Reproduced with permission from Heifets et al. 2008 (PNAS vol 105, p 10250)]

Hippocampal I-LTD shares several features in common with I-LTD in the amygdala. First, I-LTD is a heterosynaptic form of plasticity, requiring mGluR-I activation during induction (Chevaleyre and Castillo 2003) (Fig. 11.1a, b). In fact, transient application of the mGluR-I agonist DHPG is also sufficient to induce a persistent decrease in GABA release, as supported by PPR and failure rate changes. As expected, DHPG-LTD is CB1R-dependent and occludes subsequent induction of I-LTD by synaptic stimulation (Chevaleyre and Castillo 2003). Secondly, chelating intracellular calcium with BAPTA also has no effect on I-LTD in the hippocampus but blocking G-protein signaling with GDPβS in the postsynaptic cell abolishes I-LTD. Thirdly, both forms of I-LTD are dependent on PKA activity and mice lacking the presynaptic, active zone protein RIM1α, a putative substrate of PKA modulation, do not express I-LTD in either brain region (Chevaleyre et al. 2007).

While RIM1α is critical for I-LTD, it does not appear to play a role in DSI, suggesting that different molecular players are recruited in eCB-mediated STD and LTD. One factor critical for converting STD to LTD may be the duration of CB1R activation. While the maintenance of hippocampal I-LTD is CB1R-independent, a ~10-min period of CB1R activation post tetanus delivery is critical to trigger I-LTD (Chevaleyre and Castillo 2003). Recently, it was shown that during this critical period, presynaptic integration of CB1R activation *and* calcium increase arising from interneuron activity may consolidate I-LTD (Heifets et al. 2008) (Figs. 11.1c and 11.2b). This dual requirement, as proposed for excitatory synapses (Castillo and Khodakhah 2006; Singla et al. 2007), likely provides input-specificity such that only those inputs active during CB1R activation undergo eCB-mediated I-LTD (for a review, see Heifets and Castillo 2009).

Some differences between I-LTD in the hippocampus and amygdala bear mentioning here. First, inhibition of PLC and DGL activity abolishes I-LTD in the hippocampus but not in the amygdala, suggesting that the eCBs mediating I-LTD in these two regions differ (Fig. 11.1b). Secondly, in contrast to I-LTD in the amygdala, blocking postsynaptic PKA activity in CA1 pyramidal cells has no effect on hippocampal I-LTD although this I-LTD is dependent on PKA activity and most likely, on a reduction in presynaptic PKA activity (Chevaleyre et al. 2007). Stimulation protocols that are effective in triggering I-LTD in the hippocampus, ranging from moderate frequency stimulation (MFS) to high frequency stimulation (HFS) as well as the more physiologically relevant theta burst stimulation (TBS), were not tested in the amygdala. The ability of such protocols to induce I-LTD may be relevant in the context of neuronal excitability and information flow within circuits, given that HFS and TBS also triggers E-LTP in most brain regions.

Concurrent induction of E-LTP and eCB-mediated I-LTD results in an increase in the probability of a fixed postsynaptic excitatory response (EPSP) to evoke spiking (E-S coupling potentiation) (Chevaleyre and Castillo 2003), consistent with earlier work associating long-lasting disinhibition to E-S coupling potentiation in E-LTP (Abraham et al. 1987; Chavez-Noriega et al. 1989; Lu et al. 2000). Additionally, by removing inhibition, eCBs may modulate the inducibility of other forms of synaptic plasticity that are not dependent on eCBs *per se*, a phenomenon known as metaplasticity. Indeed, selective induction of I-LTD (i.e. in the absence of E-LTP induction) has been shown to prime E-LTP at nearby glutamatergic inputs by bringing weak synaptic stimuli that are normally subthreshold to threshold (Chevaleyre and Castillo 2004). Another interesting property of eCB signaling during induction of I-LTD is that eCB spread occurs within a 10 μm radius, resulting in I-LTD of a narrow set of GABAergic inputs. Such limited region of action likely confers input-specificity to E-LTP priming by I-LTD.

Dorsal Striatum

The striatum is the largest nucleus in the basal ganglia, which comprises a collection of interconnected nuclei implicated in the planning and execution of controlled motor movement, and in reward-driven motor learning (Graybiel 1995; Packard and Knowlton 2002; Schultz et al. 2003; Balleine et al. 2007). The striatum mainly consists of GABAergic medium spiny neurons (MSNs), and activity-dependent modifications of synaptic inputs to these neurons, including those triggered by eCBs, are believed to underlie adaptations in the behaviors mediated by the dorsal striatum (Jog et al. 1999; Pisani et al. 2005; Kreitzer and Malenka 2008). eCB-mediated E-LTD of cortical excitatory inputs to MSNs of the striatum was reported first (Gerdeman et al. 2002; Kreitzer and Malenka 2005, 2007) followed recently by I-LTD (Adermark and Lovinger 2009). Compared to excitatory synapses, LTD of inhibitory synapses appears to be easier to induce, presumably due to the higher expression of CB1Rs at striatal GABAergic terminals (Uchigashima et al. 2007). This notion is supported by several recent observations (Adermark and Lovinger 2009). First, while a brief LFS protocol selectively induces striatal I-LTD, induction of E-LTD typically requires relatively stronger induction protocols (e.g. pairing postsynaptic depolarization with bursts of HFS). Second, the CB1R agonist WIN at a submaximal dose (50 nM) suppresses inhibitory responses more strongly (50%) than excitatory responses (13%). Third, transient CB1R activation at a higher concentration (1 μM WIN) is sufficient to trigger persistent depression of inhibition but not excitation, even though the acute suppression by WIN at this concentration is comparable between synapse types. Lastly, combining postsynaptic loading of AEA with low frequency paired-pulse stimulation of presynaptic afferents triggers LTD of inhibitory but not excitatory transmission.

There are some interesting features of the I-LTD induced by loading AEA; some distinguishing it from eCB-mediated I-LTD in other regions. While this AEA-LTD is dependent on protein synthesis, the acute suppression of inhibitory responses by AEA is not (Adermark and Lovinger 2009), reinforcing the idea that eCB-mediated STD and LTD may involve different cellular processes. Protein synthesis is most likely required at the presynaptic site, downstream of CB1R activation, since loading AEA bypasses the eCB production step. It remains to be tested whether striatal I-LTD requires intracellular calcium rise. Interestingly, transient application of an L-type calcium channel agonist can mimic LFS-induced I-LTD (Adermark et al. 2009), suggesting that unlike I-LTD in the amygdala and hippocampus, induction of striatal I-LTD may require postsynaptic calcium influx. However, the role of L-type calcium channels in LFS-induced I-LTD needs to be directly examined with selective antagonists for these channels. Given that the level of presynaptic calcium may dictate whether I-LTD occurs, at least in the hippocampus (Heifets et al. 2008), a presynaptic role for L-type calcium channels in the induction of I-LTD cannot be discarded. Regardless of the mechanism, I-LTD may be relevant in the context of network function since it is associated with an increase in MSN excitability (Adermark and Lovinger 2009).

Ventral Tegmental Area

The VTA is a key component in the brain reward circuitry. Dopamine is a major modulator of neuronal function in the VTA, and dopaminergic neurons –the main cellular component of this brain area– are primary targets for many drugs of abuse (Carlezon and Nestler 2002; Kauer 2004; Self 2004; Di Matteo et al. 2007), including cannabinoids (Gardner 2005). Consistent with this notion, recent studies have shown that eCB-mediated I-LTD in the VTA is dependent on the activation of dopaminergic receptors (Pan et al. 2008a, b). When type 2-like dopamine receptors (D2Rs) are activated in the presence of low doses of cocaine, a blocker of the dopamine transporter, or the specific D2R agonist quinpirole, prolonged MFS (3,000 pulses at 10 Hz) of synaptic afferents triggers I-LTD. Consistent with the essential role of D2Rs, MFS-induced I-LTD in the VTA is blocked by the D2R antagonist sulpiride. While the requirement for D2Rs in I-LTD has only been tested in the VTA*, these receptors might also be involved in striatal I-LTD given the strong modulatory role of dopamine in excitatory synaptic transmission (Calabresi et al. 2007; Surmeier et al. 2007) and in the induction of eCB-mediated E-LTD in the striatum (Kreitzer and Malenka 2005).

Building on the in vivo observation that pharmacological activation of D2Rs in the striatum results in increased AEA levels (Giuffrida et al. 1999; Centonze et al. 2004), it has been postulated that D2Rs, by promoting eCB mobilization, participate in the induction of eCB-mediated LTD (Yin and Lovinger 2006). However, given that eCB mobilization is sensitive to postsynaptic neuronal activity, the possibility that in vivo AEA increases may indirectly result from D2R-mediated disinhibition, thus enhancing neuronal excitability, cannot be ruled out. Like CB1Rs, D2Rs are coupled to $G_{i/o}$ proteins and down-regulate the cAMP-PKA transduction cascade. Moreover, activation of D2Rs with higher doses of cocaine or quinpirole acutely depresses inhibitory responses in the VTA, presumably via a PKA-dependent reduction in GABA release as assessed by miniature IPSC analysis (Pan et al. 2008b). Thus, for a better understanding of the mechanisms of D2R action in

---

*During the editorial process, a report of eCB-mediated LTD at GABAergic synapses in the prefrontal cortex appeared (Chiu et al., 2010).

I-LTD, it will be important to determine whether activation of D2Rs directly leads to eCB production.

As in other eCB-mediated forms of LTD, mGluR-I activation is necessary and sufficient to trigger I-LTD in the VTA (Pan et al. 2008a). In addition, similar to hippocampal I-LTD, I-LTD in the VTA is sensitive to inhibitors of PLC, DGL and G-protein signaling, supporting a role for 2-AG. Intriguingly, I-LTD in the VTA is not accompanied by a change in PPR or coefficient of variance, which would be expected for a presynaptic form of plasticity. However, analysis of asynchronous GABA release by substituting calcium with strontium reveals changes in both frequency and amplitude of inhibitory quantal responses, suggesting that I-LTD could be expressed at both pre and postsynaptic sites. While a reduction in transmitter release is the common expression mechanism of eCB-mediated forms of LTD, the precise contribution of eCB signaling to the postsynaptic component of I-LTD in the VTA remains to be elucidated.

Importantly, I-LTD in the VTA may underlie the reduction in GABAergic transmission that is observed following repeated cocaine exposure (Pan et al. 2008a). Consistent with this possibility, in vivo administration of drugs that block I-LTD (e.g. sulpiride, AM 251) prior to each intraperitoneal cocaine injection, prevented the decrease in GABA transmission, whereas pre-treatment with a D1 receptor antagonist, which has no effect on I-LTD in the VTA, did not abolish the cocaine-associated reduction of GABAergic transmission. Furthermore, in vivo cocaine exposure occluded subsequent I-LTD ex vivo. Thus, I-LTD in the VTA may be physiologically relevant in the context of cocaine sensitization and addiction.

Common and Differential Properties of I-LTD Across Brain Regions

As we examine I-LTD across brain regions, some commonalities and differences in the underlying mechanisms emerge. In all cases, eCB-mediated I-LTD is heterosynaptically induced (Fig. 11.1a), requiring repetitive activity of nearby glutamatergic synapses and the subsequent activation of postsynaptic mGluR-I, which is necessary for eCB mobilization. Although intracellular calcium rise is essential for DSI, and eCB-mediated E-LTD, it is reportedly dispensable for the induction of I-LTD (Chevaleyre and Castillo 2003; Azad et al. 2004; Pan et al. 2008a). This disparity between the role of postsynaptic calcium in DSI and I-LTD suggests that factors in addition to CB1R activation may be required for I-LTD. Indeed, CB1R activation with the agonist WIN is insufficient by itself to trigger LTD of inhibitory transmission (Heifets et al. 2008; Pan et al. 2008a; Adermark and Lovinger 2009). Moreover, a recurring theme in eCB-LTD, including I-LTD, is the need for prolonged CB1R activation, on the order of minutes (Chevaleyre and Castillo 2003; Pan et al. 2008a; Adermark and Lovinger 2009). An activation requirement at this time scale could reflect the involvement of a slow associative signal. In the hippocampus, calcium elevations in GABAergic terminals during induction may act as this signal to regulate, in concert with eCB signaling, the balance of presynaptic kinase (i.e. PKA) and phosphatase (e.g. calcineurin) activity (Heifets et al. 2008). An obvious target of these kinases and phosphatases is the active zone protein RIM1α, which can be phosphorylated by PKA and is essential for I-LTD both in the hippocampus and amygdala (Chevaleyre et al. 2007). However, a recent study has demonstrated that while RIM1α is required for presynaptic, PKA-dependent forms of long-term plasticity, including I-LTD, direct phosphorylation of RIM1α is not required (Kaeser et al. 2008). The role of other potential targets of PKA/phosphatases (Seino and Shibasaki 2005) in eCB-mediated LTD remains to be examined.

While all known forms of I-LTD are initiated by glutamatergic synaptic activity, the identity of the eCB downstream of mGluR-I activation may vary in different brain regions. 2-AG most likely mediates I-LTD in the hippocampus and VTA, a conclusion largely supported by their sensitivity to inhibitors of DGL (Chevaleyre and Castillo 2003; Pan et al. 2008a), whereas AEA probably underlies I-LTD in the amygdala (Azad et al. 2004) and striatum (Di Marzo et al. 1994; Giuffrida et al. 1999). Given that synthesis and degradation of 2-AG and AEA differ (Piomelli 2003; Ligresti et al. 2005), the mechanisms underlying I-LTD induction in these areas may also diverge. LFS induces both amygdalar and striatal I-LTD, while slightly higher stimulation frequencies trigger I-LTD in the VTA and hippocampus. Although the requirement for D2Rs in I-LTD has only been tested in the VTA, a similar requirement may apply for striatal E-LTD where activation of D2Rs facilitates induction (Kreitzer and Malenka 2005) presumably by promoting AEA mobilization. Interestingly, 5 Hz stimulation of synaptic afferents in the prefrontal cortex triggers I-LTD when D2Rs are activated by the agonist quinpirole or by inhibiting dopamine degradation (Chiu et al. 2010). Despite the differences in eCB identity and induction protocol, I-LTD expression in the hippocampus and amygdala is mechanistically similar (Chevaleyre et al. 2007). Experimental evidence in these structures support the notion that multiple pathways leading to eCB release may converge on a generalized mechanism of I-LTD expression at the presynaptic GABAergic terminal, involving PKA signaling and the release machinery.

Physiological Relevance of Endocannabinoid-Mediated I-LTD

eCB-mediated I-LTD has been shown to enhance the excitatory influence of glutamatergic inputs on action potential generation (Chevaleyre and Castillo 2003; Adermark and Lovinger 2009) and to modulate the modifiability of excitatory synapses (Azad et al. 2004; Chevaleyre and Castillo 2004). Thus, by mediating long-term disinhibition of neurons, eCB signaling can shift the excitatory/inhibitory balance towards excitation and consequently promote signal propagation within and across neural networks. Given the importance of excitatory synaptic transmission and plasticity in experience-dependent processes (Martin et al. 2000; Malenka and Bear 2004), such wide-spread effects of eCB signaling on excitation will undoubtedly have significant impact on learning and memory. Indeed, in the intact animal, amygdalar I-LTD has been associated with extinction of learned fear (Marsicano et al. 2002) and I-LTD in the VTA has been implicated in cocaine-induced synaptic changes that may underlie drug addiction (Pan et al. 2008a).

It should be noted that the eCB system is ubiquitous in the brain, influencing both excitatory and inhibitory transmission. Hence, blocking CB1R activation with conventional pharmacological or genetic manipulations non-selectively disrupts eCB-mediated plasticity at both glutamatergic and GABAergic synapses and in many brain regions. To ascertain the physiological significance of particular forms of eCB-mediated synaptic plasticity, genetic and molecular strategies to more precisely control the eCB system in a region or cell-type specific manner will be needed. Transgenic mice lacking CB1Rs exclusively in glutamatergic, GABAergic or dopaminergic cells have been generated (Marsicano et al. 2003; Monory et al. 2007). Findings from these animals have begun to unravel the role of specific neuronal populations in the behaviorally complex actions of cannabis and eCBs (Monory et al. 2006; Monory et al. 2007; Azad et al. 2008; Puighermanal et al. 2009). In particular, exogenous activation of CB1Rs on GABAergic interneurons, but not principal neurons in the forebrain, disrupts hippocampal-dependent learning in vivo (Puighermanal et al. 2009) and inhibits LTD of excitatory synapses in the amygdala in vitro (Azad et al. 2008). Thus, CB1Rs at GABAergic synapses play a role in modulating synaptic plasticity and may underlie certain learning and memory deficits. However, the specific contribution of eCB-mediated long-term plasticity at GABAergic synapses to these physiological processes is unclear.

Although significant progress has been made in elucidating the molecular mechanisms of I-LTD induction and expression, the field of eCB research in the context of learning and memory is still in its infancy. Future experiments will be needed to answer challenging questions. For example, do induction protocols commonly employed to evoke I-LTD in vitro actually occur in vivo? Can eCB-mediated I-LTD be triggered in vivo? Given that eCB signaling affects both excitatory and inhibitory transmission, what is the net effect of eCB-mediated synaptic plasticity on network activity and therefore function? Is eCB-mediated depression at glutamatergic and GABAergic synapses differentially regulated? In the coming years, given the rapid advances in technology, uncovering and understanding the physiological role of eCB-mediated I-LTD will be in our grasp.

Acknowledgments Supported by NIH/NIDA and by NARSAD. We thank Thomas Younts for critical reading of the manuscript.

References

Abraham WC, Gustafsson B, Wigstrom H (1987) Long-term potentiation involves enhanced synaptic excitation relative to synaptic inhibition in guinea-pig hippocampus. J Physiol 394:367–380.

Adermark L, Lovinger DM (2009) Frequency-dependent inversion of net striatal output by endocannabinoid-dependent plasticity at different synaptic inputs. J Neurosci 29:1375–1380.

Adermark L, Talani G, Lovinger DM (2009) Endocannabinoid-dependent plasticity at GABAergic and glutamatergic synapses in the striatum is regulated by synaptic activity. Eur J Neurosci 29:32–41.

Aigner TG (1988) Delta-9-tetrahydrocannabinol impairs visual recognition memory but not discrimination learning in rhesus monkeys. Psychopharmacology (Berl) 95:507–511.

Alger BE (2002) Retrograde signaling in the regulation of synaptic transmission: focus on endocannabinoids. Prog Neurobiol 68:247–286.

Azad SC, Kurz J, Marsicano G, Lutz B, Zieglgansberger W, Rammes G (2008) Activation of CB1 specifically located on GABAergic interneurons inhibits LTD in the lateral amygdala. Learn Mem 15:143–152.

Azad SC, Monory K, Marsicano G, Cravatt BF, Lutz B, Zieglgansberger W, Rammes G (2004) Circuitry for associative plasticity in the amygdala involves endocannabinoid signaling. J Neurosci 24:9953–9961.

Balleine BW, Delgado MR, Hikosaka O (2007) The role of the dorsal striatum in reward and decision-making. J Neurosci 27:8161–8165.

Barnes CA (1988) Spatial learning and memory processes: the search for their neurobiological mechanisms in the rat. Trends Neurosci 11:163–169.

Beltramo M, Stella N, Calignano A, Lin SY, Makriyannis A, Piomelli D (1997) Functional role of high-affinity anandamide transport, as revealed by selective inhibition. Science 277:1094–1097.

Bender VA, Bender KJ, Brasier DJ, Feldman DE (2006) Two coincidence detectors for spike timing-dependent plasticity in somatosensory cortex. J Neurosci 26:4166–4177.

Best AR, Regehr WG (2008) Serotonin evokes endocannabinoid release and retrogradely suppresses excitatory synapses. J Neurosci 28:6508–6515.

Bisogno T, MacCarrone M, De Petrocellis L, Jarrahian A, Finazzi-Agro A, Hillard C, Di Marzo V (2001) The uptake by cells of 2-arachidonoylglycerol, an endogenous agonist of cannabinoid receptors. Eur J Biochem 268:1982–1989.

Bodor AL, Katona I, Nyiri G, Mackie K, Ledent C, Hajos N, Freund TF (2005) Endocannabinoid signaling in rat somatosensory cortex: laminar differences and involvement of specific interneuron types. J Neurosci 25:6845–6856.

Brown SP, Brenowitz SD, Regehr WG (2003) Brief presynaptic bursts evoke synapse-specific retrograde inhibition mediated by endogenous cannabinoids. Nat Neurosci 6:1048–1057.

Calabresi P, Picconi B, Tozzi A, Di Filippo M (2007) Dopamine-mediated regulation of corticostriatal synaptic plasticity. Trends Neurosci 30:211–219.

Carlezon WA, Jr., Nestler EJ (2002) Elevated levels of GluR1 in the midbrain: a trigger for sensitization to drugs of abuse? Trends Neurosci 25:610–615.

Castillo PE, Khodakhah K (2006) Biochemical confinements without walls in aspiny neurons. Nat Neurosci 9:719–720.

Centonze D, Battista N, Rossi S, Mercuri NB, Finazzi-Agro A, Bernardi G, Calabresi P, Maccarrone M (2004) A critical interaction between dopamine D2 receptors and endocannabinoids mediates the effects of cocaine on striatal gabaergic Transmission. Neuropsychopharmacology 29:1488–1497.

Chavez-Noriega LE, Bliss TV, Halliwell JV (1989) The EPSP-spike (E-S) component of long-term potentiation in the rat hippocampal slice is modulated by GABAergic but not cholinergic mechanisms. Neurosci Lett 104:58–64.

Chevaleyre V, Castillo PE (2003) Heterosynaptic LTD of hippocampal GABAergic synapses: a novel role of endocannabinoids in regulating excitability. Neuron 38:461–472.

Chevaleyre V, Castillo PE (2004) Endocannabinoid-mediated metaplasticity in the hippocampus. Neuron 43:871–881.

Chevaleyre V, Takahashi KA, Castillo PE (2006) Endocannabinoid-mediated synaptic plasticity in the CNS. Annu Rev Neurosci 29:37–76.

Chevaleyre V, Heifets BD, Kaeser PS, Sudhof TC, Castillo PE (2007) Endocannabinoid-mediated long-term plasticity requires cAMP/PKA signaling and RIM1alpha. Neuron 54:801–812.

Chiu CQ, Puente N, Grandes P, Castillo PE (2010) Dopaminergic modulation of endocannabinoid-mediated plasticity at GABAergic synapses in the prefrontal cortex. J Neurosci. 30:7236–7248.
Crozier RA, Wang Y, Liu CH, Bear MF (2007) Deprivation-induced synaptic depression by distinct mechanisms in different layers of mouse visual cortex. Proc Natl Acad Sci USA 104:1383–1388.
Devane WA, Hanus L, Breuer A, Pertwee RG, Stevenson LA, Griffin G, Gibson D, Mandelbaum A, Etinger A, Mechoulam R (1992) Isolation and structure of a brain constituent that binds to the cannabinoid receptor. Science 258:1946–1949.
Di Marzo V, Fontana A, Cadas H, Schinelli S, Cimino G, Schwartz JC, Piomelli D (1994) Formation and inactivation of endogenous cannabinoid anandamide in central neurons. Nature 372:686–691.
Di Matteo V, Pierucci M, Di Giovanni G, Benigno A, Esposito E (2007) The neurobiological bases for the pharmacotherapy of nicotine addiction. Curr Pharm Des 13:1269–1284.
Edwards DA, Kim J, Alger BE (2006) Multiple mechanisms of endocannabinoid response initiation in hippocampus. J Neurophysiol 95:67–75.
Egertova M, Giang DK, Cravatt BF, Elphick MR (1998) A new perspective on cannabinoid signalling: complementary localization of fatty acid amide hydrolase and the CB1 receptor in rat brain. Proc Biol Sci 265:2081–2085.
Ehrlich I, Humeau Y, Grenier F, Ciocchi S, Herry C, Luthi A (2009) Amygdala inhibitory circuits and the control of fear memory. Neuron 62:757–771.
Ferrari F, Ottani A, Vivoli R, Giuliani D (1999) Learning impairment produced in rats by the cannabinoid agonist HU 210 in a water-maze task. Pharmacol Biochem Behav 64:555–561.
Freund TF, Katona I, Piomelli D (2003) Role of endogenous cannabinoids in synaptic signaling. Physiol Rev 83:1017–1066.
Galante M, Diana MA (2004) Group I metabotropic glutamate receptors inhibit GABA release at interneuron-Purkinje cell synapses through endocannabinoid production. J Neurosci 24:4865–4874.
Gardner EL (2005) Endocannabinoid signaling system and brain reward: emphasis on dopamine. Pharmacol Biochem Behav 81:263–284.
Gerdeman GL, Ronesi J, Lovinger DM (2002) Postsynaptic endocannabinoid release is critical to long-term depression in the striatum. Nat Neurosci 5:446–451.
Giuffrida A, Parsons LH, Kerr TM, Rodriguez de Fonseca F, Navarro M, Piomelli D (1999) Dopamine activation of endogenous cannabinoid signaling in dorsal striatum. Nat Neurosci 2:358–363.
Graybiel AM (1995) Building action repertoires: memory and learning functions of the basal ganglia. Curr Opin Neurobiol 5:733–741.
Hampson RE, Deadwyler SA (1998) Role of cannabinoid receptors in memory storage. Neurobiol Dis 5:474–482.
Hashimotodani Y, Ohno-Shosaku T, Kano M (2007) Ca(2+)-assisted receptor-driven endocannabinoid release: mechanisms that associate presynaptic and postsynaptic activities. Curr Opin Neurobiol 17:360–365.
Heifets BD, Castillo PE (2009) Endocannabinoid signaling and long-term synaptic plasticity. Annu Rev Physiol 71:283–306.
Heifets BD, Chevaleyre V, Castillo PE (2008) Interneuron activity controls endocannabinoid-mediated presynaptic plasticity through calcineurin. Proc Natl Acad Sci USA 105:10250–10255.
Henneberger C, Redman SJ, Grantyn R (2007) Cortical efferent control of subcortical sensory neurons by synaptic disinhibition. Cereb Cortex 17:2039–2049.
Herkenham M, Lynn AB, Little MD, Johnson MR, Melvin LS, de Costa BR, Rice KC (1990) Cannabinoid receptor localization in brain. Proc Natl Acad Sci USA 87:1932–1936.
Howlett AC (2002) The cannabinoid receptors. Prostaglandins Other Lipid Mediat 68–69:619–631.
Howlett AC, Breivogel CS, Childers SR, Deadwyler SA, Hampson RE, Porrino LJ (2004) Cannabinoid physiology and pharmacology: 30 years of progress. Neuropharmacology 47 Suppl 1:345–358.

Jog MS, Kubota Y, Connolly CI, Hillegaart V, Graybiel AM (1999) Building neural representations of habits. Science 286:1745–1749.
Kaeser PS, Kwon HB, Blundell J, Chevaleyre V, Morishita W, Malenka RC, Powell CM, Castillo PE, Sudhof TC (2008) RIM1alpha phosphorylation at serine-413 by protein kinase A is not required for presynaptic long-term plasticity or learning. Proc Natl Acad Sci USA 105:14680–14685.
Kang-Park MH, Wilson WA, Kuhn CM, Moore SD, Swartzwelder HS (2007) Differential sensitivity of GABA A receptor-mediated IPSCs to cannabinoids in hippocampal slices from adolescent and adult rats. J Neurophysiol 98:1223–1230.
Kano M, Ohno-Shosaku T, Hashimotodani Y, Uchigashima M, Watanabe M (2009) Endocannabinoid-mediated control of synaptic transmission. Physiol Rev 89:309–380.
Katona I, Sperlagh B, Sik A, Kafalvi A, Vizi ES, Mackie K, Freund TF (1999) Presynaptically located CB1 cannabinoid receptors regulate GABA release from axon terminals of specific hippocampal interneurons. J Neurosci 19:4544–4558.
Kauer JA (2004) Learning mechanisms in addiction: synaptic plasticity in the ventral tegmental area as a result of exposure to drugs of abuse. Annu Rev Physiol 66:447–475.
Kreitzer AC, Malenka RC (2005) Dopamine modulation of state-dependent endocannabinoid release and long-term depression in the striatum. J Neurosci 25:10537–10545.
Kreitzer AC, Malenka RC (2007) Endocannabinoid-mediated rescue of striatal LTD and motor deficits in Parkinson's disease models. Nature 445:643–647.
Kreitzer AC, Malenka RC (2008) Striatal plasticity and basal ganglia circuit function. Neuron 60:543–554.
Lafourcade M, Elezgarai I, Mato S, Bakiri Y, Grandes P, Manzoni OJ (2007) Molecular components and functions of the endocannabinoid system in mouse prefrontal cortex. PLoS One 2:e709.
Lichtman AH, Dimen KR, Martin BR (1995) Systemic or intrahippocampal cannabinoid administration impairs spatial memory in rats. Psychopharmacology (Berl) 119:282–290.
Ligresti A, Cascio MG, Di Marzo V (2005) Endocannabinoid metabolic pathways and enzymes. Curr Drug Targets CNS Neurol Disord 4:615–623.
Lovinger DM (2008) Presynaptic modulation by endocannabinoids. Handb Exp Pharmacol: 435–477.
Lu YM, Mansuy IM, Kandel ER, Roder J (2000) Calcineurin-mediated LTD of GABAergic inhibition underlies the increased excitability of CA1 neurons associated with LTP. Neuron 26:197–205.
Maejima T, Hashimoto K, Yoshida T, Aiba A, Kano M (2001) Presynaptic inhibition caused by retrograde signal from metabotropic glutamate to cannabinoid receptors. Neuron 31:463–475.
Malenka RC, Bear MF (2004) LTP and LTD: an embarrassment of riches. Neuron 44:5–21.
Maren S, Quirk GJ (2004) Neuronal signalling of fear memory. Nat Rev Neurosci 5:844–852.
Marsicano G, Wotjak CT, Azad SC, Bisogno T, Rammes G, Cascio MG, Hermann H, Tang J, Hofmann C, Zieglgansberger W, Di Marzo V, Lutz B (2002) The endogenous cannabinoid system controls extinction of aversive memories. Nature 418:530–534.
Marsicano G, Goodenough S, Monory K, Hermann H, Eder M, Cannich A, Azad SC, Cascio MG, Gutierrez SO, van der Stelt M, Lopez-Rodriguez ML, Casanova E, Schutz G, Zieglgansberger W, Di Marzo V, Behl C, Lutz B (2003) CB1 cannabinoid receptors and on-demand defense against excitotoxicity. Science 302:84–88.
Martin SJ, Grimwood PD, Morris RG (2000) Synaptic plasticity and memory: an evaluation of the hypothesis. Annu Rev Neurosci 23:649–711.
Matsuda LA, Lolait SJ, Brownstein MJ, Young AC, Bonner TI (1990) Structure of a cannabinoid receptor and functional expression of the cloned cDNA. Nature 346:561–564.
Mechoulam R, Ben-Shabat S, Hanus L, Ligumsky M, Kaminski NE, Schatz AR, Gopher A, Almog S, Martin BR, Compton DR, et al. (1995) Identification of an endogenous 2-monoglyceride, present in canine gut, that binds to cannabinoid receptors. Biochem Pharmacol 50:83–90.

Milner B, Penfield W (1955) The effect of hippocampal lesions on recent memory. Trans Am Neurol Assoc:42–48.

Monory K, Blaudzun H, Massa F, Kaiser N, Lemberger T, Schutz G, Wotjak CT, Lutz B, Marsicano G (2007) Genetic dissection of behavioural and autonomic effects of Delta(9)-tetrahydrocannabinol in mice. PLoS Biol 5:e269.

Monory K, Massa F, Egertova M, Eder M, Blaudzun H, Westenbroek R, Kelsch W, Jacob W, Marsch R, Ekker M, Long J, Rubenstein JL, Goebbels S, Nave KA, During M, Klugmann M, Wolfel B, Dodt HU, Zieglgansberger W, Wotjak CT, Mackie K, Elphick MR, Marsicano G, Lutz B (2006) The endocannabinoid system controls key epileptogenic circuits in the hippocampus. Neuron 51:455–466.

Munro S, Thomas KL, Abu-Shaar M (1993) Molecular characterization of a peripheral receptor for cannabinoids. Nature 365:61–65.

Myers KM, Davis M (2007) Mechanisms of fear extinction. Mol Psychiatry 12:120–150.

Nevian T, Sakmann B (2006) Spine Ca2+ signaling in spike-timing-dependent plasticity. J Neurosci 26:11001–11013.

O'Keefe J (1999) Do hippocampal pyramidal cells signal non-spatial as well as spatial information? Hippocampus 9:352–364.

Packard MG, Knowlton BJ (2002) Learning and memory functions of the Basal Ganglia. Annu Rev Neurosci 25:563–593.

Pan B, Hillard CJ, Liu QS (2008a) Endocannabinoid signaling mediates cocaine-induced inhibitory synaptic plasticity in midbrain dopamine neurons. J Neurosci 28:1385–1397.

Pan B, Hillard CJ, Liu QS (2008b) D2 dopamine receptor activation facilitates endocannabinoid-mediated long-term synaptic depression of GABAergic synaptic transmission in midbrain dopamine neurons via cAMP-protein kinase A signaling. J Neurosci 28:14018–14030.

Penzo MA, Pena JL (2009) Endocannabinoid-mediated long-term depression in the avian midbrain expressed presynaptically and postsynaptically. J Neurosci 29:4131–4139.

Piomelli D (2003) The molecular logic of endocannabinoid signalling. Nat Rev Neurosci 4:873–884.

Pisani A, Centonze D, Bernardi G, Calabresi P (2005) Striatal synaptic plasticity: implications for motor learning and Parkinson's disease. Mov Disord 20:395–402.

Puighermanal E, Marsicano G, Busquets-Garcia A, Lutz B, Maldonado R, Ozaita A (2009) Cannabinoid modulation of hippocampal long-term memory is mediated by mTOR signaling. Nat Neurosci 12:1152–1158.

Ranganathan M, D'Souza DC (2006) The acute effects of cannabinoids on memory in humans: a review. Psychopharmacology (Berl) 188:425–444.

Robbe D, Kopf M, Remaury A, Bockaert J, Manzoni OJ (2002) Endogenous cannabinoids mediate long-term synaptic depression in the nucleus accumbens. Proc Natl Acad Sci USA 99:8384–8388.

Rodrigues SM, Schafe GE, LeDoux JE (2004) Molecular mechanisms underlying emotional learning and memory in the lateral amygdala. Neuron 44:75–91.

Safo PK, Regehr WG (2005) Endocannabinoids control the induction of cerebellar LTD. Neuron 48:647–659.

Schultz W, Tremblay L, Hollerman JR (2003) Changes in behavior-related neuronal activity in the striatum during learning. Trends Neurosci 26:321–328.

Seino S, Shibasaki T (2005) PKA-dependent and PKA-independent pathways for cAMP-regulated exocytosis. Physiol Rev 85:1303–1342.

Self DW (2004) Regulation of drug-taking and -seeking behaviors by neuroadaptations in the mesolimbic dopamine system. Neuropharmacology 47 Suppl 1:242–255.

Singla S, Kreitzer AC, Malenka RC (2007) Mechanisms for synapse specificity during striatal long-term depression. J Neurosci 27:5260–5264.

Sjostrom PJ, Turrigiano GG, Nelson SB (2003) Neocortical LTD via coincident activation of presynaptic NMDA and cannabinoid receptors. Neuron 39:641–654.

Smart D, Gunthorpe MJ, Jerman JC, Nasir S, Gray J, Muir AI, Chambers JK, Randall AD, Davis JB (2000) The endogenous lipid anandamide is a full agonist at the human vanilloid receptor (hVR1). Br J Pharmacol 129:227–230.

Soler-Llavina GJ, Sabatini BL (2006) Synapse-specific plasticity and compartmentalized signaling in cerebellar stellate cells. Nat Neurosci 9:798–806.

Sugiura T, Kobayashi Y, Oka S, Waku K (2002) Biosynthesis and degradation of anandamide and 2-arachidonoylglycerol and their possible physiological significance. Prostaglandins Leukot Essent Fatty Acids 66:173–192.

Sugiura T, Kondo S, Sukagawa A, Nakane S, Shinoda A, Itoh K, Yamashita A, Waku K (1995) 2-Arachidonoylglycerol: a possible endogenous cannabinoid receptor ligand in brain. Biochem Biophys Res Commun 215:89–97.

Sullivan JM (2000) Cellular and molecular mechanisms underlying learning and memory impairments produced by cannabinoids. Learn Mem 7:132–139.

Surmeier DJ, Ding J, Day M, Wang Z, Shen W (2007) D1 and D2 dopamine-receptor modulation of striatal glutamatergic signaling in striatal medium spiny neurons. Trends Neurosci 30:228–235.

Suzuki A, Josselyn SA, Frankland PW, Masushige S, Silva AJ, Kida S (2004) Memory reconsolidation and extinction have distinct temporal and biochemical signatures. J Neurosci 24:4787–4795.

Tzounopoulos T, Rubio ME, Keen JE, Trussell LO (2007) Coactivation of pre- and postsynaptic signaling mechanisms determines cell-specific spike-timing-dependent plasticity. Neuron 54:291–301.

Uchigashima M, Narushima M, Fukaya M, Katona I, Kano M, Watanabe M (2007) Subcellular arrangement of molecules for 2-arachidonoyl-glycerol-mediated retrograde signaling and its physiological contribution to synaptic modulation in the striatum. J Neurosci 27:3663–3676.

Varvel SA, Lichtman AH (2002) Evaluation of CB1 receptor knockout mice in the Morris water maze. J Pharmacol Exp Ther 301:915–924.

Varvel SA, Hamm RJ, Martin BR, Lichtman AH (2001) Differential effects of delta 9-THC on spatial reference and working memory in mice. Psychopharmacology (Berl) 157:142–150.

Wilson RI, Kunos G, Nicoll RA (2001) Presynaptic specificity of endocannabinoid signaling in the hippocampus. Neuron 31:453–462.

Yasuda H, Huang Y, Tsumoto T (2008) Regulation of excitability and plasticity by endocannabinoids and PKA in developing hippocampus. Proc Natl Acad Sci USA 105:3106–3111.

Yin HH, Lovinger DM (2006) Frequency-specific and D2 receptor-mediated inhibition of glutamate release by retrograde endocannabinoid signaling. Proc Natl Acad Sci USA 103:8251–8256.

Zhu PJ, Lovinger DM (2007) Persistent synaptic activity produces long-lasting enhancement of endocannabinoid modulation and alters long-term synaptic plasticity. J Neurophysiol 97:4386–4389.

Zimmerberg B, Glick SD, Jarvik ME (1971) Impairment of recent memory by marihuana and THC in rhesus monkeys. Nature 233:343–345.

Zygmunt PM, Petersson J, Andersson DA, Chuang H, Sorgard M, Di Marzo V, Julius D, Hogestatt ED (1999) Vanilloid receptors on sensory nerves mediate the vasodilator action of anandamide. Nature 400:452–457.

Chapter 12
Inhibitory Synaptic Plasticity and Neurotrophins

Jean-Luc Gaiarsa, Nicola Kuczewski, and Christophe Porcher

Abstract Synaptic plasticity represents the capacity of individual synapses to adjust their strength in response to modifications in the level or pattern of intrinsic or sensory-driven activity. Due to the involvement of the inhibitory transmitter GABA in brain development, synaptic function and cognition, interest in GABAergic synaptic plasticity has intensified in recent years. Neurotrophins are a family of secreted proteins that promote survival, growth and differentiation of neurons in the central nervous system. Recent advances show that besides this trophic function, neurotrophins, and brain-derived neurotrophic factor (BDNF) in particular, can exert both short-term and long-term modulations on the strength and development of GABAergic synaptic transmission. In this review, we will summarize recent findings showing that BDNF can operate as a locally released feedback modulator of GABAergic synaptic transmission.

Introduction

All brain functions, such as sensory perception, motor control and learning depend on the proper neuronal integration of excitatory and inhibitory synaptic inputs that are mediated mainly by glutamate and γ-aminobutyric acid (GABA) respectively. These synaptic inputs modify their strength and/or number over time and in response to experience in both the developing and adult central nervous system. These activity-dependent short-term and long-term changes in synaptic efficacy, known as synaptic plasticity, in turn have important effects on brain function, neuronal development and on some pathological conditions. Considering the ubiquitous distribution

J.-L. Gaiarsa (✉)
INSERM U901 (Institut National de la Santé et de la Recherche Médicale Unité 901), INMED (Institut de Neurobiologie de la Méditerranée), BP 13, Marseille Cedex 9, 13273, France
and
Aix Marseille Université, Faculté des Sciences, Marseille 13000, France
e-mail: gaiarsa@inmed.univ-mrs.fr

M.A. Woodin and A. Maffei (eds.), *Inhibitory Synaptic Plasticity*, DOI 10.1007/978-1-4419-6978-1_12, of inhibitory synapses and their role in shaping individual and population activity, information on inhibitory synaptic plasticity is mandatory to fully understand how activity-dependent changes in synaptic efficacy contribute to brain development and function in concert with plasticity at glutamatergic synapses.

The role of brain-derived neurotrophic factor (BDNF) in GABAergic synaptic plasticity has attracted particular interest. BDNF is a secreted neuronal protein, targeted to the dendrites and axon, whose expression and release are modulated by neuronal activity. Besides its classical role in supporting neuronal survival, BDNF has been reported to play a role in several key steps necessary for the full development of inhibitory GABAergic circuitry, and to mediate some of the effects of neuronal activity on GABAergic interneurons. Among these effects, BDNF has emerged as a major factor in the induction of some forms of activity-dependent GABAergic synaptic plasticity in the developing brain. Here, we first provide a brief overview of BDNF synthesis and secretion, and then summarize the recent findings showing that BDNF acts as a target-derived messenger for activity-dependent synaptic plasticity and development of inhibitory synapses in the central nervous system.

Synthesis, Targeting, Release and Cleavage of Neurotrophins

Neurotrophin Transduction Pathway

Neurotrophins are a family of secreted proteins that promote survival, growth and differentiation of neurons in the central nervous system. All members of the family - nerve growth factor (NGF), brain derived neurotrophic factor (BDNF), neurotrophin 3 (NT-3) and 4/5 (NT-4/5) - mediate their actions through two unrelated types of transmembrane receptors, the tropomyosin-related kinase receptors (Trk-Rs) and the p75 pan-neurotrophin receptor ($p75^{NTR}$), a member of the tumor necrosis factor superfamily. Unlike the non selective $p75^{NTR}$ which equally binds all neurotrophins, each Trk-R selectively binds a different neurotrophin -TrkA binds NGF, TrkB binds BDNF and NT4/5 and TrkC binds NT3. Neurotrophin binding to Trk-Rs leads to the autophosphorylation of tyrosine residues within the intracellular domains of the receptor, creating docking sites for intracellular proteins. There are three principal intracellular second messenger pathways activated by the Trk-R signaling: the Ras-Raf-ERK cascade, the phosphatidylinositol-3-OH kinase (PI3K)/Akt pathway and activation of phospholipase Cγ (PLCγ). The first two cascades lead to the activation of genes involved in cellular differentiation and survival. The activation of PLCγ leads to the production of diacylglycerol and inositol triphosphate (IP_3), which in turn triggers the release of Ca^{2+} from internal stores and activates protein kinase C (PKC) (Kaplan and Miller 2000; Huang and Reichardt 2003). PLCγ activation following BDNF binding to TrkB-R also triggers Ca^{2+} influx though TRPC channels (Amaral et al. 2007). BDNF binding and subsequent phosphorylation of TrkB-Rs has also been reported to rapidly gate a Na^+ channel, leading to postsynaptic membrane depolarization and subsequent activation of voltage-dependent Ca^{2+} channels (Kovalchuk et al. 2002). $p75^{NTR}$ lacks intrinsic catalytic activity, but signals through a series of protein-protein interactions mediated by its intracellular domain (Kaplan and Miller 2000; Lu et al. 2005). p75^{NTR} contributes to cell survival by activating the NF-κB pathway, to neurite outgrowth by regulating the RhoA pathway, and triggers apoptotic cell death by increasing ceramide production and activating the Jun kinase pathway.

Targeting and Release

BDNF is widely expressed in most brain structures and is abundant in the hippocampus, amygdala and cerebellum. Cell type analysis shows that it is expressed mainly by the principal glutamatergic neurons, and is absent from interneurons and astroglial cells, although astroglial expression of BDNF can be found in cell cultures (Ernfors et al. 1990; Wetmore et al. 1990; Lindholm et al. 1992; Gorba and Wahle 1999). Activated microglia can also express and release BDNF (Batchelor et al. 1999; Coull et al. 2005). BDNF is mainly produced in the soma, but a local dendritic synthesis occurs under certain circumstances. Transcription of mouse BDNF is controlled by at least nine distinct promoters resulting in at least 18 unique BDNF transcripts (Pruunsild et al. 2007). It has been proposed that one possible function of such complex genomic organization might be to maintain basal level of BDNF for cell survival under the control of activity-independent promoters, and when needed, to induce activity-dependent expression of BDNF for synaptic plasticity and development (Lu 2003).

After production, BDNF is sorted into the regulated and constitutive secretory pathways through the trans-Golgi network and transported to release sites in dendrites and axons. Regulated secretion of BDNF depends on sufficient elevation of intracellular Ca^{2+} concentration that is then translated into activation of CamKII (Lessmann et al. 2003; Lessmann and Brigadski 2009) (Fig. 12.1a). In addition to a rise in intracellular Ca^{2+}, basal levels of PKA are needed (Kolarow et al. 2007). This gating effect of PKA for BDNF release might reflect PKA-dependent docking of secretory granules at release sites. Regulated activity-dependent release of BDNF has been implicated in multiple processes that are important for synaptic development and plasticity in the central nervous system. Because BDNF is a sticky molecule with limited diffusion capacity (Biffo et al. 1995; Horch and Katz 2002) this neurotrophin should be released locally to act on nearby targets. This implies that multiple synaptic triggers of BDNF secretion co-exist and depending on the pattern of activity, locally secreted BDNF will exert a selective control on the development of different subpopulations of synapses and neurons. Accordingly both dendritic and axonal release of BDNF have been demonstrated (Kohara et al. 2001; Lessmann et al. 2003). Moreover several different synaptic signals leading to Ca^{2+}-dependent dendritic release of BDNF have been directly identified in neuronal cultures (Kuczewski et al. 2009) (Fig. 12.1a): (1) tetanic stimulation of presynaptic glutamatergic fibers and subsequent activation of ionotropic glutamatergic receptors (Hartmann et al. 2001), (2) prolonged depolarization of the postsynaptic neurons (Magby et al. 2006), and (3) action potentials that propagate backwards into the

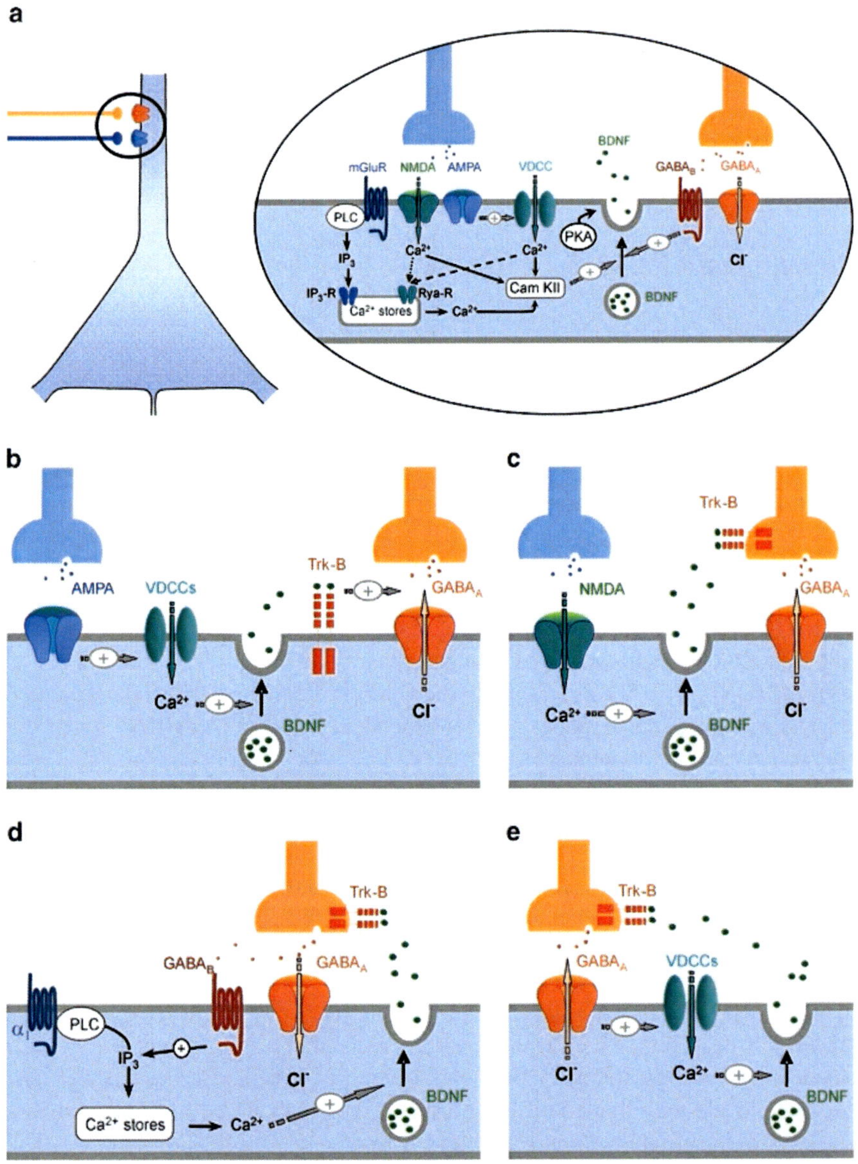

Fig. 12.1 Activity-dependent secretion of BDNF and GABAergic synaptic plasticity. (**a**) Scheme of the signaling involved in postsynaptic regulated BDNF secretion. BDNF secretion require a postsynaptic rise in intracellular Ca^{2+} concentration. This Ca^{2+} rise can result from an influx through voltage-dependent Ca^{2+} channels (VDCCs) or NMDA receptors upon membrane depolarization, or from activation of internal Ca^{2+} stores (IP3-R) following the activation of metabotropic glutamatergic receptors (mGluRs). The initial Ca^{2+} rise can be amplified by Ca^{2+}- induced Ca^{2+} secretion via ryanodine receptors (Rya-Rs). Ca^{2+} activates CaMKII leading to the fusion of the secretory granules. Basal levels of PKA activation "gate" BDNF secretion. Activation of metabotropic GABA$_B$ receptors also triggers a secretion of BDNF through a yet unknown mechanism.

dendrites (Kuczewski et al. 2008b). The activation of metabotropic glutamatergic (Canossa et al. 2001) and GABAergic (Fiorentino et al. 2009) receptors also triggers BDNF release. Depending on the location of the release site and/or stimuli used, several sources of Ca^{2+} contribute to BDNF release: NMDA channels, voltage-dependent Ca^{2+} channels (VDCCs) and intracellular Ca^{2+} stores (Fig. 12.1a). Such diversity in the Ca^{2+} source leading to BDNF secretion again supports the idea that multiple modalities of neuronal BDNF secretion can coexist.

Proneurotrophin Release and Action

Neurotrophins are synthesized as large precursor molecules (pro-neurotrophins), which are further processed within the intracellular protein transport pathway and subsequently cleaved to generate the mature forms (Lessmann et al. 2003; Lessmann and Brigadski 2009). Until recently, only the mature neurotrophins were considered biologically active, while the only function of the prodomain was to guide the sorting and targeting of the mature form. This view was recently challenged by findings that exogenously applied non-cleavable forms of proBDNF and proNGF induce biological effects that are often opposite to those of the mature forms (Lu et al. 2005). ProBDNF, acting on $p75^{NTR}$ triggers neuronal apoptosis (Lee et al. 2001), and induces long-term depression of adult glutamatergic synapses (Woo et al. 2005) and developing neuromuscular synapses (Yang et al. 2009a), while BDNF, acting on TrkB-Rs, promotes cell survival (Huang and Reichardt 2001), induces long-term potentiation of glutamatergic synapses (Baranes et al. 1998; Pang et al. 2004) and promotes the maturation of developing neuromuscular synapses (Wang et al. 1995).

Fig. 12.1 (continued) (**b–e**) BDNF-dependent long-term potentiation of $GABA_A$ receptor-mediated synaptic transmission (LTP_{GABA-A}). (**b**) In the developing rat hippocampus, ongoing glutamatergic activity induces a postsynaptic Ca^{2+} -dependent secretion of BDNF from CA3 pyramidal neurons. BDNF acting on postsynaptic TrkB-Rs triggers LTP_{GABA-A}. Whether pre- or postsynaptic modifications are involved in LTP_{GABA-A} expression is not known. (**c**) In the tectum of developing *Xenopus* tadpoles repetitive visual stimulation induces a BDNF-dependent LTP_{GABA-A}. BDNF is released from the postsynaptic neurons and acts on presynaptic TrkB-Rs. (**d**) In the developing visual cortex, high frequency stimulation triggers a BDNF-dependent LTP_{GABA-A}. The intracellular Ca^{2+} rise needed to induce LTP_{GABA-A} relies on the activation of $\alpha 1$ adreno-receptors leading to the formation of IP_3. This effect is facilitated by the activation of postsynaptic $GABA_B$ receptors. Once released from the postsynaptic pyramidal neuron, BDNF binds on presynaptic TrkB-Rs to trigger LTP_{GABA-A}. (**e**) In the developing rat hippocampus, a BDNF-dependent LTP_{GABA-A} is induced following paired stimulation of presynaptic GABAergic mossy fibers and postsynaptic CA3 pyramidal neurons. A $GABA_A$ receptor-mediated depolarization and activation of L-type VDCCs are required to trigger LTP_{GABA-A}. The location of the TrkB-Rs has not been identified. However, because the expression of LTP_{GABA-A} is expressed presynaptically, TrkB-Rs are likely located at presynaptic GABAergic mossy terminals

Although initially questioned (Matsumoto et al. 2008), substantial evidence supports the idea that a significant proportion of BDNF is released in the pro form and converted into mBDNF by extracellular proteases such as plasmin or metalloproteases (Yang et al. 2009b), and that released proBDNF is biologically active (Woo et al. 2005; Nagappan et al. 2009; Yang et al. 2009a). Thus the conversion of endogenously released proBDNF to mBDNF by extracellular plasmin appears to be critical for the late-phase of LTP at the mossy fiber-CA3 (Baranes et al. 1998) and Schaffer collateral-CA1 (Pang et al. 2004) pathways. More recently, endogenous proBDNF released from *Xenopus* myocytes in culture has been reported to cause synaptic depression and nerve terminal retraction at the developing neuromuscular junction (Yang et al. 2009a). Thus BDNF may function as a bidirectional activity-dependent messenger whose ultimate biological actions will depend on the proBDNF/mBDNF ratio.

Interestingly, a recent study reported that the extracellular conversion of secreted proBDNF in neuronal cultures is under the control of activity-dependent processes. Thus, Nagappan and collaborators showed that low frequency stimulation of neuronal cultures induced proBDNF accumulation in the extracellular medium, while high frequency stimulation caused extracellular mBDNF accumulation (Nagappan et al. 2009). The authors further reported that only when the extracellular, but not intracellular, convertases are inhibited, high frequency stimulation can induce proBDNF accumulation. Finally, they showed that high-frequency, but not low-frequency stimulation triggers the secretion of tissue plasminogen activator (tPA), a key protease required for extracellular conversion of proBDNF to mBDNF. Together, these findings support a model in which the secretion of proBDNF may have opposing effects on synaptic development and plasticity depending on its proteolytic conversion which is under the control of the pattern and/or level of synaptic activity generated by the neuronal network.

BDNF and GABAergic Synaptic Maturation

BDNF and GABA Maturation

The establishment of precise connections between neurons requires a sequence of highly regulated events that include migration, differentiation, growth and synaptogenesis. In vitro studies based on chronic treatment with exogenous BDNF or with drugs preventing BDNF-TrkB signaling showed that BDNF is necessary for the full development of inhibitory GABAergic circuitry. For example, BDNF stimulates the tangential migration of the medial ganglionic eminence cells that invade the neocortex to give rise to a majority of cortical GABAergic interneurons (Polleux et al. 2002). Once they have reached their final position, BDNF promotes the phenotypic differentiation (Marty et al. 1996a; Berghuis et al. 2004; Sakata et al. 2009) and dendritic growth of GABAergic interneurons in the cortex (Jin et al. 2003; Wirth et al. 2003; Kohara et al. 2003), hippocampus (Marty et al. 1996b) and neostriatum (Mizuno et al. 1994; Ventimiglia et al. 1995). BDNF also increases the density and/or the size of inhibitory

GABAergic terminals impinging upon hippocampal (Vicario-Abejon et al. 1998; Marty et al. 2000; MacLean Bolton et al. 2000; Yamada et al. 2002) and cortical (Kohara et al. 2001; Palizvan et al. 2004) pyramidal neurons. BDNF not only regulates the formation of GABAergic synapses, but their function as well. A hallmark of developing GABAergic synapses is that they depolarize and excite immature neurons through activation of $GABA_A$ receptors (Ben Ari et al. 2007). The depolarizing action of GABA at least in part relies on a delayed maturation of the neuronal K/Cl co-transporter KCC2 (Payne et al. 2003). BDNF up-regulates the expression levels of KCC2 thereby regulating the developmental switch from excitation to inhibition of GABAergic synaptic transmission (Aguado et al. 2003; Carmona et al. 2006).

Evidences that BDNF plays a crucial role in GABAergic synaptic development are found not only in tissue culture, but also from in vivo studies. For example, the development of inhibitory circuitry is impaired in BDNF (Abidin et al. 2008) and TrkB-R KO mice (Carmona et al. 2006). In these mice GABAergic markers, the number of GABAergic terminals and GABAergic synaptic activity are reduced compared to age matched wild type mice. Conversely, in embryonic or postnatal BDNF over-expressing mice the emergence of the mature GABAergic circuitry is accelerated (Huang et al. 1999; Aguado et al. 2003). In the visual cortex for instance, early over-expression of BDNF accelerates the development of GABAergic inhibition, leading to an earlier termination of the critical window for ocular dominance plasticity, which is controlled by GABAergic inhibition, and to a precocious visual acuity (Huang et al. 1999).

BDNF as Target-Derived Messenger for GABAergic Synapse Maturation

Single cell transfection has been used to demonstrate that BDNF can act as a target-derived messenger for GABAergic synaptic development. Because the secretion of BDNF remains under the control of the cell release mechanism, single cell BDNF over-expression allows creation of a defined cellular source of BDNF that more closely mimics the endogenous situation than does bath application. Single cell over-expression of BDNF in neuronal cultures results in a selective increase in presynaptic GAD65 expression and GABAergic synaptic activity in the transfected neuron (Ohba et al. 2005). This effect was mimicked by exogenous BDNF, prevented by the tyrosine kinase inhibitor, k252a, and attenuated by postsynaptic expression of a kinase-defective TrkB-R. BDNF-GFP was also expressed in a small number of neurons within hippocampal neuronal cultures obtained from *bdnf−/−* mice (Singh et al. 2006). The number of GABAergic terminals and the level of GABAergic synaptic activity were significantly increased in BDNF-expressing neurons when compared to non-transfected *bdnf−/−* neurons. Thus, even if the level of BDNF in transfected neurons is different from those produced in control neurons these findings support the idea that BDNF released from one single target neuron can promote the development of GABAergic terminals.

Whether released BDNF acts in a target-selective manner, or as a diffusible messenger, has been addressed by loss of function. A single-cell gene knock-out method was used to create a small number of BDNF-deficient neurons in organotypic slice cultures of visual cortex (Kohara et al. 2007). The authors then loaded one single GABAergic neuron with biocytin to estimate the number of GABAergic terminals on two postsynaptic neurons innervated by the same presynaptic interneurons. The number of biocytin-labeled and GAD65-positive boutons established on the BDNF-KO neurons was reduced compared to the neighboring control neurons. The frequency of miniature GABAergic synaptic currents recorded in BDNF-KO neurons was reduced as well. Thus, postsynaptic release of endogenous BDNF can promote the formation of GABAergic synapses in a highly specific targeted manner.

Whether constitutive or regulated release of BDNF in GABAergic maturation is involved has also been addressed. Several studies have revealed that the trophic action of BDNF on GABAergic circuitry is linked to synaptic activity generated by the neuronal network. In the cerebellum, hippocampus and cortex, activity-dependent enhancement of GABAergic synaptic activity can be mimicked by chronic treatment with exogenous BDNF or prevented by interfering with BDNF-TrkB signaling (Seil and Drake-Baumann 1994; Rutherford et al. 1997; Marty et al. 2000; Seil and Drake-Baumann 2000). Conversely, chronic silencing of the neuronal network impaired the maturation of GABAergic interneurons and synapses, an effect reversed by exogenous application of BDNF. Moreover, mice in which the activity-dependent promoter of BDNF transcription (promoter IV) has been silenced exhibited a significant deficit in cortical GABAergic circuitry (Hong et al. 2008; Sakata et al. 2009). Altogether, these observations show that activity-dependent release of BDNF plays a role in the development of the inhibitory circuits. Interestingly, $GABA_A$ and $GABA_B$ receptor activation up-regulates the expression of BDNF in developing hypothalamic (Obrietan et al. 2002) and hippocampal neuronal cultures (Berninger et al. 1995; Ghorbel et al. 2005). Thus, a positive feedback loop between GABA and BDNF exists early in development, where GABA stimulates BDNF expression in glutamatergic neurons and BDNF released from these neurons facilitates the development of inhibitory synapses.

BDNF and GABAergic Synaptic Efficacy

Acute Effects on GABAergic Synaptic Transmission

The observation that BDNF can be released from target neurons to promote the development of GABAergic synapses has prompted investigators to examine the effect of BDNF on GABAergic synaptic transmission. The first study on the effect of neurotrophins in CNS synapses showed that NT-3 induced a rapid increase in the excitability of cultured cortical neurons due to an acute suppression of GABAergic synaptic transmission (Kim et al. 1994). Since then, the abundant BDNF literature has revealed a multitude of effects from which it is often difficult to establish a clear picture. Both pre- and post-synaptic modifications have been reported. In the rat hippocampus, BDNF modulates paired-pulse ratio (PPR) and coefficient of variation (CV) of evoked inhibitory postsynaptic currents (IPSCs) (Frerking et al. 1998; Sivakumaran et al. 2009), the number of failures of evoked minimal IPCS (Sivakumaran et al. 2009) and the frequency of miniature IPSCs (Bolton et al. 2000; Wardle and Poo 2003; Gubellini et al. 2005) suggesting a presynaptic site of action. Other studies showed that BDNF reduces the amplitude of postsynaptic $GABA_A$ receptor-mediated responses to applied GABA (Tanaka et al. 1997; Brunig et al. 2001), down-regulates the surface expression of $GABA_A$-Rs (Brunig et al. 2001; Cheng and Yeh 2003), and modulates the phosphorylation state of the $GABA_A$-Rs (Jovanovic et al. 2004). BDNF also modulates the efficacy of GABAergic synaptic responses acting on the neuronal K^+-Cl^- co-transporter KCC2, thus regulating intracellular chloride homeostasis (Rivera et al. 2002; Wardle and Poo 2003).

To add to this complexity, the effects are often bidirectional and depending on the age, brain regions or cell type both potentiation and depression of GABAergic synaptic transmission have been observed. For example, BDNF potentiates GABAergic synaptic transmission in hippocampal CA1 pyramidal neurons isolated from postnatal day (P) six rats, while the same amount of BDNF inhibits GABAergic synaptic transmission at P14 (Mizoguchi et al. 2003a). Similarly BDNF up-regulates KCC2 expression in the developing brain (Aguado et al. 2003) while down-regulation is observed in adult neurons (Rivera et al. 2002). Cell specific effects of BDNF also exist. In cerebellar slices, BDNF enhances GABA responses in Purkinje cells, but attenuates those recorded from granule cells (Cheng and Yeh 2005). In cultured hippocampal neurons, BDNF decreases the amplitude of evoked IPSCs recorded from glutamatergic neurons, while increasing IPSC amplitude in GABAergic neurons from the same cultures (Wardle and Poo 2003).

One possible explanation for these bidirectional effects is the recruitment of different intracellular signaling cascades. In neuronal cultures, BDNF-induced potentiation and depression both require an elevation of intracellular Ca^{2+} in response to TrkB-R and PLC activation (Mizoguchi et al. 2003a, b; Cheng and Yeh 2005). However, activation of CaMKII is necessary for the potentiation of $GABA_A$ responses by BDNF, but not for their inhibition (Mizoguchi et al. 2003a). Similarly, BDNF recruits different transduction signaling pathways to control the phosphorylation state of the β3 subunit of the $GABA_A$ receptor, namely PKC-mediated phosphorylation and PP2A-mediated dephosphorylation (Jovanovic et al. 2004; Kanematsu et al. 2006). Therefore, BDNF provides a large range of possibilities for the regulation of GABAergic synaptic transmission in developing and adult brain structures.

Long-Term Effects on GABAergic Synaptic Transmission

Although the studies mentioned above provide important information on the actions of BDNF, it has yet to be shown that endogenous native BDNF is able to affect GABAergic synaptic transmission. Moreover, because BDNF must be secreted to mediate its effects, crucial questions remain to be addressed in terms of the physiological pattern of synaptic activity that triggers the release of BDNF, as well as the impact this release may have on network development.

The first study suggesting a contribution of endogenous neurotrophins to long-term changes in inhibitory synaptic transmission was obtained by Kotak and collaborators (Kotak et al. 2001). The authors showed that inhibitory projections in the developing gerbil auditory brain stem undergo long-term depression (LTD) when activated at low frequency. This LTD was prevented by the alkaloid tyrosine kinase inhibitor k252a and mimicked by bath applied NT-3. Since then, a long-term potentiation of $GABA_A$ receptor-mediated synaptic transmission (LTP_{GABA-A}) has been observed in the developing rat hippocampus upon repeated postsynaptic firing of CA3 pyramidal neurons (Gubellini et al. 2005) or following paired stimulation of pre- and postsynaptic neurons, a plasticity known as spike timing dependent plasticity (STDP) (Sivakumaran et al. 2009), in the developing visual cortex following high frequency stimulation (HFS) (Inagaki et al. 2008), and in the tectum of developing *Xenopus* tadpoles after repetitive visual stimulation (Liu et al. 2007) (Fig. 12.1b–e).

In all of these cases, a postsynaptic rise in intracellular Ca^{2+} concentration ($[Ca^{2+}]_i$) is required to shape the strength of inhibitory synapses (Fig. 12.1b–e) as indicated by the observation that loading the postsynaptic neuron with the Ca^{2+} chelator BAPTA prevents the induction of GABAergic plasticity. The $[Ca^{2+}]_i$ rise results from the opening of postsynaptic L-type VDCCs in the developing rat hippocampus (Gubellini et al. 2005; Sivakumaran et al. 2009) (Fig. 12.1b, e), from the opening of postsynaptic NMDA channel in the developing *Xenopus* tectum (Liu et al. 2007) (Fig. 12.1c), and from a release of Ca^{2+} from postsynaptic internal stores following PLC activation and IP_3 formation in the developing rat visual cortex (Komatsu 1996) (Fig. 12.1d). In addition to the postsynaptic $[Ca^{2+}]_i$ rise, the transduction pathway of GABAergic plasticity in these structures involved extracellular endogenous BDNF and the subsequent activation of TrkB-Rs. Thus, LTP_{GABA-A} was prevented by the application of an extracellular BDNF scavenger (i.e. TrkB-R fused to a human IgG Fc domain, TrkB-IgG (Shelton et al. 1995)) or by the alkaloid tyrosine kinase inhibitor k252a, applied in the bath before and/or during the conditioning protocol. Blocking the BDNF-TrkB signaling pathway after LTP_{GABA-A} induction however has no effect, showing that activation of TrkB-Rs is required for induction, but not maintenance (Gubellini et al. 2005; Inagaki et al. 2008). Because pyramidal neurons, but not GABAergic interneurons, produce BDNF (Ernfors et al. 1990; Gorba and Wahle 1999), the postsynaptic $[Ca^{2+}]_i$ rise may be required to trigger the release of BDNF necessary for LTP_{GABA-A}. Several experimental findings support this hypothesis. In the developing rat hippocampus, GABAergic plasticity was rescued by bath applied BDNF, under conditions that prevented the postsynaptic rise in $[Ca^{2+}]_i$ (i.e. in the presence of the L-type VDCC blocker, nifedipine (Kuczewski et al. 2008a)). Similarly, when the cells were loaded with the Ca^{2+} chelator BAPTA, HFS failed to induce LTP_{GABA-A} in the developing visual cortex. This failure was rescued by application of extracellular BDNF which has no effect on basal inhibitory synaptic transmission *per se* (Inagaki et al. 2008). These rescue experiments support a postsynaptic release of endogenous BDNF from postsynaptic neurons in response to spiking activity or HFS. Accordingly, Lessmann and collaborators reported that HFS applied to cultured hippocampal neurons led to a Ca^{2+}-dependent dendritic release of BDNF (Hartmann et al. 2001). Similarly, Kuczewski and collaborators (2008) monitored activity-dependent release of BDNF and demonstrated that spontaneous action potentials that back-propagate into the dendrites (b-APs) trigger a Ca^{2+}-dependent dendritic release of BDNF-GFP (Kuczewski et al. 2008b). Most importantly, the authors showed that the tens of b-APs induced in one single neuron are sufficient to trigger the release of endogenous BDNF and activate TrkB signaling in neighboring neurons. Thus, although appropriate procedures are required to visualize endogenous BDNF release in living tissue, these results show that conditioning protocols leading to long-term changes in GABAergic synaptic efficacy can trigger a Ca^{2+}-dependent dendritic release of BDNF.

In both developing rat hippocampus and visual cortex LTP_{GABA-A} was restricted to the stimulated GABAergic fibers (Inagaki et al. 2008; Sivakumaran et al. 2009). Why active inputs are selectively potentiated compared to less active inputs to the same neuron is presently not known. At least two possible mechanisms can be proposed. One is that during the conditioning protocol, the stimulated GABAergic inputs trigger a localized release of BDNF that does not spread to non-stimulated GABAergic inputs. Indeed, the spread of neurotrophins is limited by their sticky nature and by the presence of truncated TrkB-R lacking tyrosine kinase activity located on both neurons and glia (Biffo et al. 1995). An alternative mechanism is that presynaptic activity gates the BDNF-TrkB signaling cascade. Accordingly, in the visual cortex, BDNF potentiates GABAergic synapses that are stimulated at high frequency, but not those stimulated at a low rate (Liu et al. 2007; Inagaki et al. 2008).

How BDNF-dependent LTP_{GABA-A} is expressed has been investigated. The expression of BDNF-dependent LTP_{GABA-A} is associated with an increase in the frequency, but not amplitude, of miniature IPSCs (Caillard et al. 1999), with a decrease in the paired-pulse ratio (PPR) (Inagaki et al. 2008; Sivakumaran et al. 2009), and coefficient of variation (CV) of evoked IPSCs (Sivakumaran et al. 2009), and with a decrease in the percentage of failures of minimal evoked IPSCs (Sivakumaran et al. 2009). These results therefore suggest that BDNF acts as a retrograde-messenger released from pyramidal neurons during the conditioning protocol to modulate the level or efficacy of GABAergic synaptic transmission. The precise transduction pathway involved in this cascade remains to be elucidated, but possible candidate include an up-regulation of presynaptic VDCCs involved in transmitter release (Baldelli et al. 2005), presynaptic vesicle-associated proteins, or the number of presynaptic docked vesicles (reviewed in (Tyler et al. 2002)).

Most of the interest in synaptic plasticity stems from the possibility that it might play a role in brain function. With this in mind, the conditioning protocol used to trigger synaptic plasticity deserved particular attention. Compelling evidence suggest that physiological conditioning protocols can induce BDNF-dependent plasticity at GABAergic synapses. Early in development, ongoing synaptic activity in the developing rat hippocampus is characterized by the presence of spontaneous network-driven events, termed giant depolarizing potentials (GDPs), both in vitro (Ben-Ari et al. 1989) and in vivo (Leinekugel et al. 2002). Pairing spontaneous GDPs with presynaptic stimulation of mossy fibers (MF) potentiated MF-CA3 GABAergic synapses in neonatal (P1-P6) rat hippocampal slices (Kasyanov et al. 2004). The GDP-induced plasticity required a postsynaptic influx of Ca^{2+} through L-type VDCCs, and is likely expressed presynaptically as an increase in the probability of GABA release. The contribution of BDNF was not investigated in this study. However, the same group reported that paired stimulation of presynaptic MF and postsynaptic CA3 pyramidal neurons triggers a BDNF-dependent potentiation of MF-CA3 GABAergic synapses (Sivakumaran et al. 2009). Moreover, the same group showed that pairing GDPs with presynaptic Schaffer collateral stimulation triggers a Ca^{2+}- and BDNF-dependent potentiation of glutamatergic synaptic transmission in the developing rat CA1 hippocampal region (Mohajerani et al. 2007). In a subsequent study, Kuczewski and collaborators combined electrophysiological recordings and ELISA immuno-detection on neonatal rat hippocampal slices to show that ongoing glutamatergic synaptic activity up-regulates the secretion of BDNF through the activation of postsynaptic AMPA receptors and L-type VDCCs and uncovers a BDNF-dependent long-term potentiation of GABAergic synaptic activity (Fig. 12.1b) (Kuczewski et al. 2008a). Not only endogenous, but also sensory-driven synaptic activity can trigger a potentiation of GABAergic activity. Thus, a potentiation of GABAergic inputs to *Xenopus* tectal neurons can be induced in vivo by repetitive visual stimuli (Liu et al. 2007). This potentiation requires the activation of postsynaptic NMDA receptors, the release of BDNF and subsequent activation of presynaptic TrkB-Rs. It is also associated with presynaptic changes in GABA release (Fig. 12.1c). Thus, BDNF-dependent GABAergic plasticity can be induced by physiological patterns of endogenous or sensory-driven synaptic activity. The precise function and consequences of this plasticity are presently unknown. However, because activity-dependent release of BDNF is required for the full development of GABAergic circuits, the above studies suggest that BDNF-dependent plasticity may contribute to experience-dependent maturation of inhibition and hence to the functional development of the central nervous system.

Conclusion

The above summarized studies have established that activity-dependent secretion of BDNF can act locally to modulate the maturation and efficacy of GABAergic synapses. In future experiments, a careful morphological identification of the type of interneurons or the somato-dendritic location of the GABAergic synapses will be required to fully understand the functional implications of plasticity at inhibitory synapses. Thus, in the adult hippocampus and cortex, different interneurons target specific areas and differentially control the excitability of their target cells. Whether all type of GABAergic interneurons express a BDNF-dependent synaptic plasticity is not known. If they do, together with the fact that BDNF diffusion is restricted, this would imply that multiple synaptic triggers of BDNF secretion might coexist. Depending on the pattern of activity generated by the neuronal network, BDNF could exert a selective control on the development of different subpopulations of GABAergic neurons or synapses. Indeed, in the neonatal rat hippocampus at least two different forms of GABAergic plasticity are produced by spontaneous ongoing activity in the developing hippocampus: an associative potentiation induced by spontaneous network-driven GDPs paired with stimulation of presynaptic GABAergic mossy fiber (Kasyanov et al. 2004) and a heterosynaptic potentiation of GABAergic synapses induced by ongoing glutamatergic synaptic activity (Kuczewski et al. 2008a). The former potentiation might contribute to the fine organization of the CA3 mossy fiber connections, while the latter might be important for the formation of functionally balanced excitatory and inhibitory inputs.

Several other issues remain to be addressed. For example, although the effects of the unprocessed form of neurotrophins (pro-neurotrophins) on cell survival and glutamatergic synaptic strength have been studied (Lu et al. 2005), whether and how pro-neurotrophins affect GABAergic synaptic transmission is not known. Similarly, while many studies have studied the reciprocal interaction between developing inhibitory synapses and endogenous BNDF, much less is known about the effect of this neurotrophin on inhibitory circuits in the adult brain. Answering these questions might have important implications in the management of disorders which involve alterations in BDNF release and signaling.

References

Abidin I, Eysel UT, Lessmann V, Mittmann T (2008) Impaired GABAergic inhibition in the visual cortex of brain-derived neurotrophic factor heterozygous knockout mice. J Physiol 586: 1885–1901.

Aguado F, Carmona MA, Pozas E, Aguilo A, Martinez-Guijarro FJ, Alcantara S, Borrell V, Yuste R, Ibanez CF, Soriano E (2003) BDNF regulates sponaneous correlated activity at early developmental stages by increasing synaptogenesis and expression of K^+/Cl^- co-transport. Development 130: 1287–1280.

Amaral MD, Chapleau CA, Pozzo-Miller L (2007) Transient receptor potential channels as novel effectors of brain-derived neurotrophic factor signaling: potential implications for Rett syndrome. Pharmacol Ther 113: 394–409.

Baldelli P, Hernandez-Guijo JM, Carabelli V, Carbone E (2005) Brain-derived neurotrophic factor enhances GABA release probability and nonuniform distribution of N- and P/Q-type channels on release sites of hippocampal inhibitory synapses. J Neurosci 25: 3358–3368.

Baranes D, Lederfein D, Huang YY, Chen M, Bailey CH, Kandel ER (1998) Tissue plasminogen activator contributes to the late phase of LTP and to synaptic growth in the hippocampal mossy fiber pathway. Neuron 21: 813–825.

Batchelor PE, Liberatore GT, Wong JY, Porritt MJ, Frerichs F, Donnan GA, Howells DW (1999) Activated macrophages and microglia induce dopaminergic sprouting in the injured striatum and express brain-derived neurotrophic factor and glial cell line-derived neurotrophic factor. J Neurosci 19: 1708–1716.

Ben Ari Y, Gaiarsa JL, Tyzio R, Khazipov R (2007) GABA: a pioneer transmitter that excites immature neurons and generates primitive oscillations. Physiol Rev 87: 1215–1284.

Ben-Ari Y, Cherubini E, Corradetti R, Gaïarsa J-L (1989) Giant synaptic potentials in immature rat CA3 hippocampal neurones. J Physiol (Lond) 416: 303–325.

Berghuis P, Dobszay MB, Sousa KM, Schulte G, Mager PP, Hartig W, Gorcs TJ, Zilberter Y, Ernfors P, Harkany T (2004) Brain-derived neurotrophic factor controls functional differentiation and microcircuit formation of selectively isolated fast-spiking GABAergic interneurons. Eur J Neurosci 20: 1290–1306.

Berninger B, Marty S, Zafra F, Da Penha Berzaghi M, Thoenen H, Lindholm D (1995) GABAergic stimulation switches from enhancing to repressing BDNF expression in rat hippocampal neurons during maturation in vitro. Development 121: 2327–2335.

Biffo S, Offenhauser N, Carter BD, Barde YA (1995) Selective binding and internalisation by truncated receptors restrict the availability of BDNF during development. Development 121: 2461–2470.

Bolton MM, Lo DC, Sherwood NT (2000) Long-term regulation of excitatory and inhibitory synaptic transmission in hippocampal cultures by brain-derived neurotrophic factor. Prog Brain Res 128: 203–218.

Brunig I, Penschuck S, Berninger B, Benson J, Fritschy JM (2001) BDNF reduces miniature inhibitory postsynaptic currents by rapid downregulation of GABA(A) receptor surface expression. Eur J Neurosci 13: 1320–1328.

Caillard O, Ben-Ari Y, Gaïarsa J-L (1999) Long-term potentiation of GABAergic synaptic transmission in neonatal rat hippocampus. J Physiol (Lond) 518.1: 109–119.

Canossa M, Gartner A, Campana G, Inagaki N, Thoenen H (2001) Regulated secretion of neurotrophins by metabotropic glutamate group I (mGluRI) and Trk receptor activation is mediated via phospholipase C signalling pathways. EMBO J 20: 1640–1650.

Carmona MA, Pozas E, Martinez A, Espinosa-Parrilla JF, Soriano E, Aguado F (2006) Age-dependent spontaneous hyperexcitability and impairment of GABAergic function in the hippocampus of mice lacking trkB. Cereb Cortex 16: 47–63.

Cheng Q, Yeh HH (2003) Brain-derived neurotrophic factor attenuates mouse cerebellar granule cell GABA(A) receptor-mediated responses via postsynaptic mechanisms. J Physiol 548: 711–721.

Cheng Q, Yeh HH (2005) PLCgamma signaling underlies BDNF potentiation of Purkinje cell responses to GABA. J Neurosci Res 79: 616–627.

Coull JA, Beggs S, Boudreau D, Boivin D, Tsuda M, Inoue K, Gravel C, Salter MW, De Koninck Y (2005) BDNF from microglia causes the shift in neuronal anion gradient underlying neuropathic pain. Nature 438: 1017–1021.

Ernfors P, Wetmore C, Olson L, Persson H (1990) Identification of cells in rat brain and peripheral tissues expressing mRNA for members of the nerve growth factor family. Neuron 5: 511–526.

Fiorentino H, Kuczewski N, Diabira D, Ferrand N, Pangalos MN, Porcher C, Gaiarsa JL (2009) GABAB receptors activation triggers BDNF release and promotes the maturation of GBAergic synapses. J Neurosci in press.

Frerking M, Malenka RC, Nicoll RA (1998) Brain-derived neurotrophic factor (BDNF) modulates inhibitory, but not excitatory, transmission in the CA1 region of the hippocampus. J Neurophysiol 80: 3383–3386.

Ghorbel MT, Becker KG, Henley JM (2005) Profile of changes in gene expression in cultured hippocampal neurones evoked by the GABAB receptor agonist baclofen. Physiol Genomics 22: 93–98.

Gorba T, Wahle P (1999) Expression of TrkB and TrkC but not BDNF mRNA in neurochemically identified interneurons in rat visual cortex in vivo and in organotypic cultures. Eur J Neurosci 11: 1179–1190.

Gubellini P, Ben Ari Y, Gaiarsa JL (2005) Endogenous neurotrophins are required for the induction of GABAergic long-term potentiation in the neonatal rat hippocampus. J Neurosci 25: 5796–5802.

Hartmann M, Heumann R, Lessmann V (2001) Synaptic secretion of BDNF after high-frequency stimulation of glutamatergic synapses. EMBO J 20: 5887–5897.

Hong EJ, McCord AE, Greenberg ME (2008) A biological function for the neuronal activity-dependent component of Bdnf transcription in the development of cortical inhibition. Neuron 60: 610–624.

Horch HW, Katz LC (2002) BDNF release from single cells elicits local dendritic growth in nearby neurons. Nat Neurosci 5: 1177–1184.

Huang EJ, Reichardt LF (2001) Neurotrophins: roles in neuronal development and function. Annu Rev Neurosci 24: 677–736.

Huang EJ, Reichardt LF (2003) Trk receptors: roles in neuronal signal transduction. Annu Rev Biochem 72: 609–642.

Huang ZJ, Kirkwood A, Pizzorusso T, Porciatti V, Morales B, Bear MF, Maffei L, Tonegawa S (1999) BDNF regulates the maturation of inhibition and the critical period of plasticity in mouse visual cortex. Cell 98: 739–755.

Inagaki T, Begum T, Reza F, Horibe S, Inaba M, Yoshimura Y, Komatsu Y (2008) Brain-derived neurotrophic factor-mediated retrograde signaling required for the induction of long-term potentiation at inhibitory synapses of visual cortical pyramidal neurons. Neurosci Res 61: 192–200.

Jin X, Hu H, Mathers PH, Agmon A (2003) Brain-derived neurotrophic factor mediates activity-dependent dendritic growth in nonpyramidal neocortical interneurons in developing organotypic cultures. J Neurosci 23: 5662–5673.

Jovanovic JN, Thomas P, Kittler JT, Smart TG, Moss SJ (2004) Brain-derived neurotrophic factor modulates fast synaptic inhibition by regulating GABA(A) receptor phosphorylation, activity, and cell-surface stability. J Neurosci 24: 522–530.

Kanematsu T, Yasunaga A, Mizoguchi Y, Kuratani A, Kittler JT, Jovanovic JN, Takenaka K, Nakayama KI, Fukami K, Takenawa T, Moss SJ, Nabekura J, Hirata M (2006) Modulation of GABA(A) receptor phosphorylation and membrane trafficking by phospholipase C-related inactive protein/protein phosphatase 1 and 2A signaling complex underlying brain-derived neurotrophic factor-dependent regulation of GABAergic inhibition. J Biol Chem 281: 22180–22189.

Kaplan DR, Miller FD (2000) Neurotrophin signal transduction in the nervous system. Curr Opin Neurobiol 10: 381–391.

Kasyanov AM, Safiulina VF, Voronin LL, Cherubini E (2004) GABA-mediated giant depolarizing potentials as coincidence detectors for enhancing synaptic efficacy in the developing hippocampus. Proc Natl Acad Sci USA 101: 3967–3972.

Kim HG, Wang T, Olafsson P, Lu B (1994) Neurotrophin 3 potentiates neuronal activity and inhibits gamma-aminobutyratergic synaptic transmission in cortical neurons. Proc Natl Acad Sci USA 91: 12341–12345.

Kohara K, Kitamura A, Adachi N, Nishida M, Itami C, Nakamura S, Tsumoto T (2003) Inhibitory but not excitatory cortical neurons require presynaptic brain-derived neurotrophic factor for dendritic development, as revealed by chimera cell culture. J Neurosci 23: 6123–6131.

Kohara K, Kitamura A, Morishima M, Tsumoto T (2001) Activity-dependent transfer of brain-derived neurotrophic factor to postsynaptic neurons. Science 291: 2419–2423.

Kohara K, Yasuda H, Huang Y, Adachi N, Sohya K, Tsumoto T (2007) A local reduction in cortical GABAergic synapses after a loss of endogenous brain-derived neurotrophic factor, as revealed by single-cell gene knock-out method. J Neurosci 27: 7234–7244.

Kolarow R, Brigadski T, Lessmann V (2007) Postsynaptic secretion of BDNF and NT-3 from hippocampal neurons depends on calcium calmodulin kinase II signaling and proceeds via delayed fusion pore opening. J Neurosci 27: 10350–10364.

Komatsu Y (1996) $GABA_B$ receptors, monoamine receptors, and postsynaptic inositol trisphosphate-induced Ca^{2+} release are involved in the induction of long-term potentiation at visual cortical inhibitory synapses. J Neurosci 16: 6342–6352.

Kotak VC, Dimattina C, Sanes DH (2001) GABAB and Trk receptor signaling mediates long-lasting inhibitory synaptic depression. J Neurophysiol 86: 536–540.

Kovalchuk Y, Hanse E, Kafitz KW, Konnerth A (2002) Postsynaptic Induction of BDNF-Mediated Long-Term Potentiation. Science 295: 1729–1734.

Kuczewski N, Langlois A, Fiorentino H, Bonnet S, Marissal T, Diabira D, Ferrand N, Porcher C, Gaiarsa JL (2008a) Spontaneous glutamatergic activity induces a BDNF-dependent potentiation of GABAergic synapses in the newborn rat hippocampus. J Physiol 586: 5119–5128.

Kuczewski N, Porcher C, Ferrand N, Fiorentino H, Pellegrino C, Kolarow R, Lessmann V, Medina I, Gaiarsa JL (2008b) Backpropagating action potentials trigger dendritic release of BDNF during spontaneous network activity. J Neurosci 28: 7013–7023.

Kuczewski N, Porcher C, Lessmann V, Medina I, Gaiarsa JL (2009) Activity-Dependent Dendritic Release of BDNF and Biological Consequences. Mol Neurobiol 39: 37–49.

Lee R, Kermani P, Teng KK, Hempstead BL (2001) Regulation of cell survival by secreted proneurotrophins. Science 294: 1945–1948.

Leinekugel X, Khazipov R, Cannon R, Hirase H, Ben-Ari Y, Buzsáki G (2002) Correlated burst of activity in the neonatal rat hippocmpus in vivo. Science 296: 2049–2052.

Lessmann V, Brigadski T (2009) Mechanisms, locations, and kinetics of synaptic BDNF secretion: An update. Neurosci Res.

Lessmann V, Gottmann K, Malcangio M (2003) Neurotrophin secretion: current facts and future prospects. Prog Neurobiol 69: 341–374.

Lindholm D, Castren E, Hengerer B, Zafra F, Berninger B, Thoenen H (1992) Differential Regulation of Nerve Growth Factor (NGF) Synthesis in Neurons and Astrocytes by Glucocorticoid Hormones. Eur J Neurosci 4: 404–410.

Liu Y, Zhang LI, Tao HW (2007) Heterosynaptic scaling of developing GABAergic synapses: dependence on glutamatergic input and developmental stage. J Neurosci 27: 5301–5312.

Lu B (2003) BDNF and activity-dependent synaptic modulation. Learn Mem 10: 86–98.

Lu B, Pang PT, Woo NH (2005) The yin and yang of neurotrophin action. Nat Rev Neurosci 6: 603–614.

MacLean Bolton M, Pittman AJ, Lo DC (2000) Brain-derived neurotrophic factor differentially regulatesexcitatory and inhibitory synaptic transmission in hippoampal neurons. J Neurosci 20: 3221–3232.

Magby JP, Bi C, Chen ZY, Lee FS, Plummer MR (2006) Single-cell characterization of retrograde signaling by brain-derived neurotrophic factor. J Neurosci 26: 13531–13536.

Marty S, Berninger B, Carroll P, Thoenen H (1996a) GABAergic stimulation regulates the phenotype of hippocampal interneurons through the regulation of brain-derived neurotrophic factor. Neuron 16: 565–570.

Marty S, Carroll P, Cellerino A, Castren E, Staiger V, Thoenen H, Lindholm D (1996b) Brain-derived neurotrophic factor promotes the differentiation of various hippocampal nonpyramidal neurons, including Cajal-Retzius cells, in organotypic slice cultures. J Neurosci 16: 675–687.

Marty S, Wehrlé R, Sotelo C (2000) Neuronal activity and brain-derived neurotrophic factor regulate the density of inhibitory synapses in organotypic slice cultures of postnatal hippocampus. J Neurosci 20: 8087–8095.

Matsumoto T, Rauskolb S, Polack M, Klose J, Kolbeck R, Korte M, Barde YA (2008) Biosynthesis and processing of endogenous BDNF: CNS neurons store and secrete BDNF, not pro-BDNF. Nat Neurosci 11: 131–133.

Mizoguchi Y, Ishibashi H, Nabekura J (2003a) The action of BDNF on GABA(A) currents changes from potentiating to suppressing during maturation of rat hippocampal CA1 pyramidal neurons. J Physiol 548: 703–709.

Mizoguchi Y, Kanematsu T, Hirata M, Nabekura J (2003b) A rapid increase in the total number of cell surface functional GABAA receptors induced by brain-derived neurotrophic factor in rat visual cortex. J Biol Chem 278: 44097–44102.

Mizuno K, Carnahan J, Nawa H (1994) Brain-derived neurotrophic factor promotes differentiation of striatal GABAergic neurons. Dev Biol 165: 243–256.

Mohajerani MH, Sivakumaran S, Zacchi P, Aguilera P, Cherubini E (2007) Correlated network activity enhances synaptic efficacy via BDNF and the ERK pathway at immature CA3 CA1 connections in the hippocampus. Proc Natl Acad Sci USA.

Nagappan G, Zaitsev E, Senatorov VV, Jr., Yang J, Hempstead BL, Lu B (2009) Control of extracellular cleavage of ProBDNF by high frequency neuronal activity. Proc Natl Acad Sci USA 106: 1267–1272.

Obrietan K, Gao XB, Van den Pol AN (2002) Excitatory actions of GABA increase BDNF expression via a MAPK-CREB-dependent mechanism--a positive feedback circuit in developing neurons. J Neurophysiol 88: 1005–1015.
Ohba S, Ikeda T, Ikegaya Y, Nishiyama N, Matsuki N, Yamada MK (2005) BDNF locally potentiates GABAergic presynaptic machineries: target-selective circuit inhibition. Cereb Cortex 15: 291–298.
Palizvan MR, Sohya K, Kohara K, Maruyama A, Yasuda H, Kimura F, Tsumoto T (2004) Brain-derived neurotrophic factor increases inhibitory synapses, revealed in solitary neurons cultured from rat visual cortex. Neuroscience 126: 955–966.
Pang PT, Teng HK, Zaitsev E, Woo NT, Sakata K, Zhen S, Teng KK, Yung WH, Hempstead BL, Lu B (2004) Cleavage of proBDNF by tPA/plasmin is essential for long-term hippocampal plasticity. Science 306: 487–491.
Payne JA, Rivera C, Voipio J, Kaila K (2003) Cation-chloride co-transporters in neuronal communication, development and trauma. Trends Neurosci 26: 199–206.
Polleux F, Whitford KL, Dijkhuizen PA, Vitalis T, Ghosh A (2002) Control of cortical interneuron migration by neurotrophins and PI3-kinase signaling. Development 129: 3147–3160.
Pruunsild P, Kazantseva A, Aid T, Palm K, Timmusk T (2007) Dissecting the human BDNF locus: bidirectional transcription, complex splicing, and multiple promoters. Genomics 90: 397–406.
Rivera C, Li H, Thomas-Crusells J, Lahtinen H, Viitanen T, Nanobashvili A, Kokaia Z, Airaksinen MS, Voipio J, Kaila K, Saarma M (2002) BDNF-induced TrkB activation down-regulates the K^+-Cl^- cotransporter KCC2 and impairs neuronal Cl^- extrusion. J Cell Biol 159: 747–752.
Rutherford LC, DeWan A, Lauer HM, Turrigiano GG (1997) Brain-derived neurotrophic factor mediates the activity-dependent regulation of inhibition in neocortical cultures. J Neurosci 17: 4527–4535.
Sakata K, Woo NH, Martinowich K, Greene JS, Schloesser RJ, Shen L, Lu B (2009) Critical role of promoter IV-driven BDNF transcription in GABAergic transmission and synaptic plasticity in the prefrontal cortex. Proc Natl Acad Sci USA 106: 5942–5947.
Seil FJ, Drake-Baumann R (1994) Reduced cortical inhibitory synaptogenesis in organotypic cerebellar cultures developing in the absence of neuronal activity. J Comp Neurol 342: 366–377.
Seil FJ, Drake-Baumann R (2000) TrkB receptor ligands promote activity-dependent inhibitory synaptogenesis. J Neurosci 20: 5367–5373.
Shelton DL, Sutherland J, Gripp J, Camerato T, Armanini MP, Phillips HS, Carroll K, Spencer SD, Levinson AD (1995) Human trks: molecular cloning, tissue distribution, and expression of extracellular domain immunoadhesins. J Neurosci 15: 477–491.
Singh B, Henneberger C, Betances D, Arevalo MA, Rodriguez-Tebar A, Meier JC, Grantyn R (2006) Altered balance of glutamatergic/GABAergic synaptic input and associated changes in dendrite morphology after BDNF expression in BDNF-deficient hippocampal neurons. J Neurosci 26: 7189–7200.
Sivakumaran S, Mohajerani MH, Cherubini E (2009) At immature mossy-fiber-CA3 synapses, correlated presynaptic and postsynaptic activity persistently enhances GABA release and network excitability via BDNF and cAMP-dependent PKA. J Neurosci 29: 2637–2647.
Tanaka T, Saito H, Matsuki N (1997) Inhibition of $GABA_A$ synaptic responses by brain-derived neurotrophic factor (BDNF) in rat hippocampus. J Neurosci 17: 2959–2966.
Tyler WJ, Perrett SP, Pozzo-Miller LD (2002) The role of neurotrophins in neurotransmitter release. Neuroscientist 8: 524–531.
Ventimiglia R, Mather PE, Jones BE, Lindsay RM (1995) The neurotrophins BDNF, NT-3 and NT-4/5 promote survival and morphological and biochemical differentiation of striatal neurons in vitro. Eur J Neurosci 7: 213–222.
Vicario-Abejon C, Collin C, McKay RD, Segal M (1998) Neurotrophins induce formation of functional excitatory and inhibitory synapses between cultured hippocampal neurons. J Neurosci 18: 7256–7271.
Wang T, Xie K, Lu B (1995) Neurotrophins promote maturation of developing neuromuscular synapses. J Neurosci 15: 4796–4805.
Wardle RA, Poo MM (2003) Brain-derived neurotrophic factor modulation of GABAergic synapses by postsynaptic regulation of chloride transport. J Neurosci 23: 8722–8732.

Wetmore C, Ernfors P, Persson H, Olson L (1990) Localization of brain-derived neurotrophic factor mRNA to neurons in the brain by in situ hybridization. Exp Neurol 109: 141–152.

Wirth MJ, Brun A, Grabert J, Patz S, Wahle P (2003) Accelerated dendritic development of rat cortical pyramidal cells and interneurons after biolistic transfection with BDNF and NT4/5. Development 130: 5827–5838.

Woo NH, Teng HK, Siao CJ, Chiaruttini C, Pang PT, Milner TA, Hempstead BL, Lu B (2005) Activation of p75NTR by proBDNF facilitates hippocampal long-term depression. Nat Neurosci 8: 1069–1077.

Yamada MK, Nakanishi K, Ohba S, Nakamura T, Ikegaya Y, Nishiyama N, Matsuki N (2002) Brain-derived neurotrophic factor promotes the maturation of GABAergic mechanisms in cultured hippocampal neurons. J Neurosci 22: 7580–7585.

Yang F, Je HS, Ji Y, Nagappan G, Hempstead B, Lu B (2009a) Pro-BDNF-induced synaptic depression and retraction at developing neuromuscular synapses. J Cell Biol 185: 727–741.

Yang J, Siao CJ, Nagappan G, Marinic T, Jing D, McGrath K, Chen ZY, Mark W, Tessarollo L, Lee FS, Lu B, Hempstead BL (2009b) Neuronal release of proBDNF. Nat Neurosci 12: 113–115.

Index

V